D1728564

Internal Combustion Engine Bearings Lubrication
in Hydrodynamic Bearings

*Series Editor
Piotr Breitkopf*

Internal Combustion Engine Bearings Lubrication in Hydrodynamic Bearings

Dominique Bonneau
Aurelian Fatu
Dominique Souchet

WILEY

First published 2014 in Great Britain and the United States by ISTE Ltd and John Wiley & Sons, Inc.

Apart from any fair dealing for the purposes of research or private study, or criticism or review, as permitted under the Copyright, Designs and Patents Act 1988, this publication may only be reproduced, stored or transmitted, in any form or by any means, with the prior permission in writing of the publishers, or in the case of reprographic reproduction in accordance with the terms and licenses issued by the CLA. Enquiries concerning reproduction outside these terms should be sent to the publishers at the undermentioned address:

ISTE Ltd
27-37 St George's Road
London SW19 4EU
UK

www.iste.co.uk

John Wiley & Sons, Inc.
111 River Street
Hoboken, NJ 07030
USA

www.wiley.com

© ISTE Ltd 2014
The rights of Dominique Bonneau, Aurelian Fatu and Dominique Souchet to be identified as the authors of this work have been asserted by them in accordance with the Copyright, Designs and Patents Act 1988.

Library of Congress Control Number: 2014942902

British Library Cataloguing-in-Publication Data
A CIP record for this book is available from the British Library
ISBN 978-1-84821-684-6

Printed and bound in Great Britain by CPI Group (UK) Ltd., Croydon, Surrey CR0 4YY

Contents

PREFACE ix

NOMENCLATURE. xi

CHAPTER 1. KINEMATICS AND DYNAMICS OF CRANK SHAFT–CONNECTING ROD–PISTON LINKAGE 1

   1.1. Kinematic model of crank shaft–connecting
rod–piston linkage. 2
      1.1.1. Model description 2
      1.1.2. Expressions of angular velocities 5
      1.1.3. Expressions of velocity for points A, G_2 and B. 5
      1.1.4. Expressions of connecting rod angular acceleration
and points G_2 and B accelerations. 7
   1.2. Efforts in the links between the crank shaft, the connecting rod
and the piston 8
      1.2.1. Hypothesis and data. 8
      1.2.2. Dynamics equations for the piston. 9
      1.2.3. Dynamics equations for the axis 9
      1.2.4. Dynamics equations for the connecting rod. 10
      1.2.5. Dynamics equations for the crank shaft 11
      1.2.6. Efforts for frictionless links 12
   1.3. Load diagram correction in the case of large deformations. 13
      1.3.1. Kinematics of crank shaft–connecting
rod–piston system with mobility 14
      1.3.2. Dynamics of crank shaft–connecting
rod–piston system with mobility 20

1.4. Examples of link efforts between the elements of crank
shaft–connecting rod–piston system 23
    1.4.1. Data 23
    1.4.2. Load diagrams for the connecting rod big
    end bearing 24
    1.4.3. Load diagrams for a connecting rod small
    end bearing 26
    1.4.4. Load diagrams for a crank shaft main bearing 27
    1.4.5. Engine torque 28
1.5. Bibliography 29

CHAPTER 2. THE CRANK SHAFT–CONNECTING ROD LINK 31

2.1. Geometrical and mechanical characteristics of the
connecting rod big end bearing 31
2.2. Lubricant supply 33
2.3. Correction of the load diagram in the case of
large deformations 34
2.4. Multibody models 38
    2.4.1. Interfaces and interactions: main assumptions 39
    2.4.2. Equations of unilateral contact with friction
    and equilibrium equations 41
    2.4.3. Compliance matrices 42
    2.4.4. Finite element modeling of the contact
    in the joint plane 46
    2.4.5. Modelization of the contact between the
    housing and the shells 65
2.5. Case of V engines 72
2.6. Examples of connecting rod big end bearing
computations 79
    2.6.1. Presentation of connecting rods and corresponding
    load diagrams 80
    2.6.2. Geometry and lubricant data 84
    2.6.3. Analysis of some isothermal results 85
    2.6.4. Influence of mesh downsizing 96
    2.6.5. Search of potential damage zones
    due to cavitation 98
    2.6.6. Examples taking into consideration
    thermoelastohydrodynamic effects 100
2.7. Bibliography 118

CHAPTER 3. THE CONNECTING ROD–PISTON LINK 123

3.1. Geometrical particularities and mechanics of
connecting rod–piston link 123
3.2. Lubricant supply 125
3.3. Example of computation for a connecting rod
small end bearing with the axis embedded into the piston. 127
3.4. Complete model of the connecting rod–piston link 133
    3.4.1. Equations. 134
    3.4.2. Integration of dynamics equation 137
    3.4.3. Piston structural model 139
    3.4.4. Example: the piston–axis–connecting rod
    small end link for a Formula 1 engine 142
3.5. Bibliography 158

CHAPTER 4. THE ENGINE BLOCK–CRANK SHAFT LINK 161

4.1. Geometrical and mechanical particularities
of the engine block – crank shaft link 161
4.2. Lubricant supply 162
4.3. Calculus of an isolated crank shaft bearing 163
4.4. Complete model of the engine block – crank shaft link. 170
    4.4.1. Model presentation 171
    4.4.2. Expression of the elastic deformations 173
    4.4.3. Expression of the film thickness 175
    4.4.4. Equation system 175
    4.4.5. Resolution method 178
    4.4.6. Examples. 180
4.5. Bibliography 196

**CHAPTER 5. INFLUENCE OF INPUT PARAMETERS
AND OPTIMIZATION** 197

5.1. Design of experiments method 197
5.2. Identification of the input parameters: example 201
5.3. Multiobjective optimization. 202
5.4. Optimization of a connecting rod big end
bearing: example 204
    5.4.1. Viscosity factors. 208
    5.4.2. Radial clearance factor 209
    5.4.3. Radial shape defect 209
    5.4.4. Axial shape defect. 210
    5.4.5. Shell bore relief factors 210

5.4.6. Supply pressure and temperature. 210
5.4.7. Power loss 211
5.4.8. Contact pressure velocity factor 211
5.4.9. Severity criterion based on the minimum
film thickness 212
5.4.10. Leakage 213
5.4.11. Global functioning temperature. 214
5.4.12. Bearing optimization method 214
5.5. Bibliography 221

INDEX 223

Preface

This volume is the fourth and final part of the series devoted to hydrodynamic bearings.

Volume 1 [BON 14a] describes in detail the lubricant physical properties that play an essential role in hydrodynamic phenomena, followed by hydrodynamic lubrication equations and models for their numerical solutions. Part of Volume 1 also gives "elastohydrodynamic" (EHD) model descriptions.

Volume 2 [BON 14b] is devoted to the study of mixed lubrication. The role of surface roughness is analyzed using the corresponding numerical techniques both from a hydrodynamic and a surface asperity contact point of view. This volume also addresses the issue of surface wear in the aforementioned context.

Volume 3 [BON 14c] describes several thermohydrodynamic (THD) models and thermoelastohydrodynamic (TEHD) problems. This volume ends with a description of general algorithms used in computational software designed to describe bearings under large non-stationary loads.

This last volume (Volume 4) addresses specific problems related to engine and compressor bearing calculations.

Chapter 1 of this volume describes the kinematic and dynamic relationships of the mobile part (crank shaft, connecting rod, piston) of an internal combustion engine.

Chapters 2, 3 and 4 are devoted to different system bearings, respectively, connecting rod big and small end bearings and crank shaft journal bearings. The specific problems associated with each one are analyzed in detail: lubricant supply, multibody rods and bearings with coupled operation, etc. Several examples specific to different types of combustion engines are illustrated in these three chapters.

Chapter 5 describes practical bearing calculation techniques for optimizing lubrication conditions. A reliable, simple and easy approach for optimizing multiple objects, based on methods used in experimental design, is used to create mathematical models for optimizing chosen parameter values, e.g. in the case of engine bearings: power loss and criteria for severity evaluation. These models are later used in the optimization phase to replace complicated numerical simulations used for calculating bearing lubrication parameters.

An application to internal combustion engine connecting rod big end bearing calculations is described in detail.

Bibliography

[BON 14a] BONNEAU D., FATU A., SOUCHET D., *Hydrodynamic Bearings*, ISTE, London and John Wiley & Sons, New York, 2014

[BON 14b] BONNEAU D., FATU A., SOUCHET D., *Mixed Lubrication in Hydrodynamic Bearings*, ISTE, London and John Wiley & Sons, New York, 2014

[BON 14c] BONNEAU D., FATU A., SOUCHET D., *Thermo-hydrodynamic Lubrication in Hydrodynamic Bearings*, ISTE, London and John Wiley & Sons, New York, 2014

Nomenclature

Points, basis, repairs, links and domains

A	point at the junction crank shaft – connecting rod (2D model)
B	point at the junction connecting rod – piston (2D model)
O	origin point of lubricant film repair (developed bearing)
O_c	origin point of the repair attached to the housing (bearing center)
O_a	origin point of the repair attached to the shaft
x, y, z	Cartesian basis for the film (developed bearing)
$\mathbf{x_c, y_c, z_c}$	Cartesian basis for the housing
B_i	$\{\mathbf{x_i, y_i, z_0}\}$, basis of the repair $\mathfrak{R}_i$
$\mathfrak{R}_0$	$\{O, \mathbf{x_0, y_0, z_0}\}$, engine block repair
$\mathfrak{R}_1$	$\{O, \mathbf{x_1, y_1, z_0}\}$, crank shaft repair
$\mathfrak{R}_2$	$\{A, \mathbf{x_2, y_2, z_0}\}$, connecting rod repair
$\mathfrak{R}_3$	$\{O, \mathbf{x_3, y_3, z_0}\}$, small end shaft repair
$\mathfrak{R}_4$	$\{O, \mathbf{x_4, y_4, z_0}\}$, piston repair
L_{ij}	link between the solid S_i and the solid S_j

Scalars

B	m	bearing half-width
C	m	bearing radial clearance
C_p	J kg^{-1} °C^{-1}	specific heat
D	Pa; m	universal variable representing p else $r - h$
E	N m^{-2}	Young modulus
F_N, F_T	N	normal and tangential contact force components

H	W m^{-2} °C^{-1}	thermal transfer coefficient
I_i	kg m^2	central moment of inertia for solid S_i with respect to O z axis
K_N, K_T	N m^{-1}	penalization stiffness for a contact problem
L	m	bearing width
L_b	m	connecting rod length
M_{xc}	N m	$\mathbf{x}_c$ component for the moment at O_c of $\Im_{pressure}$ torsor
M_{yc}	N m	$\mathbf{y}_c$ component for the moment at O_c of $\Im_{pressure}$ torsor
R	m	bearing radius
R_v	m	crank shaft radius
S	m^2	piston surface
U	m s^{-1}	shaft peripherical velocity for a bearing
V	m s^{-1}	squeeze velocity for a bearing
W	m s^{-1}	shaft axial velocity for a bearing
W_{xc}	N	$\mathbf{x}_c$ component of $\Im_{pressure}$ torsor resultant
W_{yc}	N	$\mathbf{y}_c$ component of $\Im_{pressure}$ torsor resultant
d	m	distance between the piston shaft and the crank shaft center (2D model)
f		Coulomb friction coefficient
h	m	lubricant film thickness
k		relative position of the con rod mass center with respect to the big end axis
k	W m^{-1} °C^{-1}	thermal conductivity
m_i	kg	mass of solid S_i
p	Pa	pressure in the lubricant film
p_{supply}	Pa	supply pressure for a bearing
p_g	Pa	combustion gas pressure acting on the piston
t	s	time
u	m s^{-1}	circumferential velocity component at a point into the film
u_N, u_T	m	normal and tangential displacements at a contact point
v	m s^{-1}	velocity squeeze component at a point into the film
w	m s^{-1}	axial velocity component at a point into the film
x	m	circumferential coordinate for a point into the film
y	m	coordinate in the thickness direction for a point into the film
z	m	axial coordinate for a point into the film
α	°Pa^{-1}	piezoviscosity coefficient

β	°C^{-1}	thermoviscosity coefficient
$\varepsilon_x, \varepsilon_y$		relative eccentricity components
ζ_x, ζ_y	rad	misalignment components for the shaft into the housing
θ	rad	angular coordinate for a film point for a bearing
θ_v	rad	crank shaft angle
μ	Pa.s	lubricant dynamic viscosity
ρ	kg m^{-3}	lubricant density
σ	m	combined roughness of film walls
φ	rad	connecting rod angle with respect to the engine block
χ	rad	con rod small end axis angle with respect to the engine block
ψ	rad	connecting rod angle with respect to the crank shaft
ω	rad s^{-1}	shaft angular velocity with respect to the housing

Dimensioned parameters

$\overline{R}$	R_v/L_b
$\overline{d}$	d/L_b
$\overline{h}$	h/σ

Torsors

$\mathfrak{S}_{pressure}$	pressure actions exerted on the housing
$\mathfrak{S}_{applied\ load}$	loading for a shaft or thrust bearing

Matrices

[A]	N Pa^{-1}	integration matrix
[C]	m Pa^{-1}	compliance matrix
[K]	N m^{-1}	stiffness matrix
[A_i]		matrix of the problem i equation discretized by the finite element method
[J]		Jacobian matrix

Indices

F	film or lubricant
S	shaft, solid
supply	lubricant supply
amb	ambient medium

Acronyms

BDC	bottom dead center
CPV	contact pressure velocity product
DOE	design of experiments
EA	evolutionary algorithm
EHD	elastohydrodynamic
GT	global thermal (method)
MFT, MOFT	minimum (oil) film thickness
MOFP	maximum oil film pressure
MTM	mean temperature method
PTM	parabolic temperature profile method
SAE	Society of Automotive Engineers
TDC	top dead center
TEHD	thermoelastohydrodynamic

1

Kinematics and Dynamics of Crank Shaft–Connecting Rod–Piston Linkage

In an internal combustion engine, the combination of mechanical parts, which allows the force exerted by the combustion of gas to be transformed into rotational movement resulting in vehicle wheel rotation, is referred to as "the moving part". This includes pistons, piston pins, connecting rod bearings, connecting rods and the crank shaft. The small elements related to sealing or assembling the piston and lock rings for piston pin positioning are not discussed. Because of their low mass, these connecting elements barely influence the forces created by the moving part. The connecting rod consists of a set of assembled solid objects: connecting rod beam, cap, bearings, screws, washers and possibly nuts. It will be assumed that there is no movement between these various elements of the connecting rod.

The aim of this chapter is to determine the kinematic relationships between elements of the moving part and forces involved in the junctions. Although these elements are made up of elastic materials and are thus capable of deformation under the effect of force transmission, in this chapter they will be considered non-deformable.

The junctions between the elements of the moving part are generally made with sliding bearings. These require a lubricant film layer in order to function well. The extra hundredths of a millimeter occupied by the layer contribute to the general junction mobility and additional mobility of very small amplitudes. This notable extra mobility significantly complicates the kinematic model of the moving part.

A study on dynamics using a complete kinematic model requires knowledge of junction static and dynamic properties, e.g. all the coefficients of stiffness and damping matrices of each link, the dimensions of each matrix is equal to the degrees of freedom. The developments and examples presented in other chapters of this book show that the elastohydrodynamic behaviors of lubricated bearings in non-stationary conditions are such that they make it impossible to construct a dynamic model where the bearings would be represented by stiffness and damping matrices known in advance. Once again, the purpose of this chapter is to give the necessary background needed for junction forces calculations. A kinematic and dynamic model where the junctions are reduced to their core mobility is sufficient for acquiring results with the desired precision. In section 1.3, a model is developed that takes into account the extra mobility added by the significant deformability of the connecting rod bearing.

1.1. Kinematic model of crank shaft–connecting rod–piston linkage

1.1.1. *Model description*

The moving part examined is assumed to be those of a single-cylinder engine made up of five non-deformable bodies numbered from 0 to 4 as follows:

– 0: engine block;

– 1: the crank shaft with the center of rotation O and the center of mass G_1;

– 2: the connecting rod AB with the center of mass G_2;

– 3: the element coupling the connecting rod and the piston;

– 4: the piston.

The mechanism plan consists of a closed kinematic chain located in a Cartesian plane O ($\mathbf{x}_0$, $\mathbf{y}_0$). Only the basic mobility of the junctions between these solid bodies will be considered as follows:

– L_{01}: pivoting link between the engine block and the crank shaft with the center O and axis O $\mathbf{z}_0$;

– L_{12}: pivoting link between the crank shaft and the connecting rod with the center A and axis A $\mathbf{z}_0$;

– L_{23}: sliding pivoting link between the connecting rod and the axis with the center B and axis B $\mathbf{z}_0$;

– L_{34}: pivoting link between the piston pin and the piston with the center B and axis B $\mathbf{z}_0$;

– L_{04}: annular linear link[1] between the engine block and the piston with the center O and axis D x_0.

Figure 1.1 shows the kinematic diagram of this model. In this figure, we can see the basis and the parameters used. To simplify notation, the basis vectors use the same notations as the frame axis. For example, the frame $\Re_0$ of the engine block has O as the point of origin and axis Ox_0, Oy_0 and Oz_0. The base of this frame consists of orthonormal vectors x_0, y_0 and z_0. Because of this similarity, the vector notation is not shown in the figure.

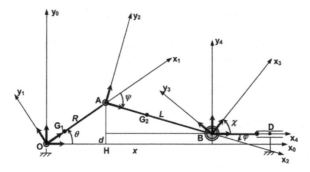

Figure 1.1. *Kinematic model*

The reference frames and basis are defined as follows:

– $\Re_0 \equiv \{O, x_0, y_0, z_0\}$: engine block frame;

– $\Re_1 \equiv \{O, x_1, y_1, z_0\}$: crank shaft frame;

– $\Re_2 \equiv \{A, x_2, y_2, z_0\}$: connecting rod frame;

– $\Re_3 \equiv \{B, x_3, y_3, z_0\}$: small end shaft frame;

– $\Re_4 \equiv \{B, x_4, y_4, z_0\} \equiv \{B, x_0, y_0, z_0\}$: piston frame;

– $B_i \equiv \{x_i, y_i, z_0\}$: basis of the frame $\Re_i$.

The geometrical values used are as follows:

– R: the radius of the crank shaft;

– a: distance from center O to the crank shaft's center of mass G_2;

– L: connecting rod length;

– $k = AG_2/L$: position relative to point G_2 of the connecting rod;

---

[1] This choice is necessary for achieving a non-hyperstatic assembly.

– *d*: default alignment between the cylinder axis and the center of the crank shaft.

The parameters used are as follows:

– θ: crank shaft angle (x_0, x_1) relative to the engine block;

– ψ: connecting rod angle (x_1, x_2) with respect to the crank shaft;

– φ: connecting rod angle (x_0, x_2) with respect to the engine block;

– χ: connecting rod small end axis angle (x_0, x_3) with respect to the engine block;

– *x*: piston pin position with respect to the center of the crank shaft.

Except for the angle χ of the piston axis that is independent, these parameters are geometrically interrelated to one another as follows:

$$HA = R\sin\theta = d - L\sin\varphi$$

$$x = R\cos\theta + L\cos\varphi$$

$$\varphi = \theta + \psi$$

Knowing that $\overline{R} = \dfrac{R}{L}$ and $\overline{d} = \dfrac{d}{L}$, the following equations can be derived:

$$\sin\varphi = \overline{d} - \overline{R}\sin\theta \qquad [1.1]$$

$$\cos\varphi = \sqrt{1 - (\overline{d} - \overline{R}\sin\theta)^2} \qquad [1.2]$$

$$\varphi = Arc\sin(\overline{d} - \overline{R}\sin\theta) \qquad [1.3]$$

$$x = L\left(\dfrac{R}{L}\cos\theta + \cos\varphi\right) = L\left[\overline{R}\cos\theta + \sqrt{1 - (\overline{d} - \overline{R}\sin\theta)^2}\right] \qquad [1.4]$$

for which only the independent parameters θ and χ are absolutely necessary.

1.1.2. Expressions of angular velocities

The crank shaft can be assumed to move at an angular velocity of $\omega = \dot{\theta}$. This velocity will be the kinematic input parameter used for studying the hydrodynamic properties of crank shaft journal bearings. For the bearing at the big end of the connecting rod, this role will be carried out by the angular velocity $\dot{\psi}$. Using the previously derived equation [1.3], the connecting rod angular velocity $\dot{\varphi}$ can be calculated as:

$$\dot{\varphi} = \frac{-\overline{R}\dot{\theta}\cos\theta}{\sqrt{1-(\overline{d}-\overline{R}\sin\theta)^2}} = \frac{-\overline{R}\dot{\theta}\cos\theta}{\cos\varphi} \qquad [1.5]$$

The angular velocity $\dot{\psi}$ can then be expressed as:

$$\dot{\psi} = \dot{\theta}\left[\frac{\overline{R}\cos\theta}{\sqrt{1-(\overline{d}-\overline{R}\sin\theta)^2}} - 1\right] = \dot{\theta}\left(\frac{\overline{R}\cos\theta}{\cos\varphi} - 1\right) \qquad [1.6]$$

The angular velocity $\dot{\chi}$ cannot be determined solely by kinematic relationships.

1.1.3. Expressions of velocity for points A, G_2 and B

The velocity relation between points of non-deformable bodies allows us to calculate the velocities of different points in the mechanism.

$$V_{01}(A) = R\dot{\theta}y_1 \qquad [1.7]$$

$$V_{02}(G_2) = R\dot{\theta}\left[y_1 - \frac{k\cos\theta}{\cos\varphi}y_2\right] \qquad [1.8]$$

$$V_{02}(B) = R\dot{\theta}(-\sin\theta + \tan\varphi\cos\theta)x_0 \qquad [1.9]$$

The vectors y_1 and y_2 can be easily replaced by their separated components in the basis B_0:

$$y_1 = -\sin\theta x_0 + \cos\theta y_0$$

$$y_2 = -\sin\varphi x_0 + \cos\varphi y_0$$

which gives:

$$V_{02}(G_2) = R\dot\theta\left[(-\sin\theta + k\tan\varphi\cos\theta)x_0 + (1-k)\cos\theta y_0\right] \quad [1.10]$$

With the help of equations [1.1] and [1.2], the velocities can be expressed as a function of a single parameter θ.

In most engines, the piston is in alignment with the center of the crank shaft, so that d becomes zero. In this case, the angular velocity $\dot\varphi$ can be written as:

$$\dot\varphi = \frac{-\overline{R}\dot\theta\cos\theta}{\sqrt{1-\overline{R}^2\sin^2\theta}} \quad [1.11]$$

and the velocity of points G_2 and B can be expressed as:

$$V_{02}(G_2) = R\dot\theta\left[-(1+k\frac{\overline{R}\cos\theta}{\sqrt{1-\overline{R}^2\sin^2\theta}})\sin\theta x_0 + (1-k)\cos\theta y_0\right] \quad [1.12]$$

$$V_{02}(B) = -R\dot\theta\sin\theta\left[1+\frac{\overline{R}\cos\theta}{\sqrt{1-\overline{R}^2\sin^2\theta}}\right]x_0 \quad [1.13]$$

1.1.4. *Expressions of connecting rod angular acceleration and points G_2 and B accelerations*

From equation [1.5], the angular acceleration $\ddot{\varphi}$ can be obtained as:

$$\ddot{\varphi} = \frac{\overline{R}}{\cos\varphi}\left[-\ddot{\theta}\cos\theta + \dot{\theta}^2\frac{\left(\cos^2\varphi\sin\theta + \overline{R}\cos^2\theta\sin\varphi\right)}{\cos^2\varphi}\right] \qquad [1.14]$$

From [3.10] and [3.9], the equations describing acceleration of points G_2 and B can be derived as:

$$\Gamma(G_2/\Re_0) = R\ddot{\theta}\left[(-\sin\theta + k\tan\varphi\cos\theta)x_0 + (1-k)\cos\theta y_0\right]$$
$$+ R\dot{\theta}^2\left[\left(-\cos\theta - k\tan\varphi\sin\theta - k\overline{R}\frac{\cos^2\theta}{\cos^3\varphi}\right)x_0 - (1-k)\sin\theta y_0\right] \qquad [1.15]$$

$$\Gamma(B/\Re_0) = R\left[\ddot{\theta}(-\sin\theta + \tan\varphi\cos\theta) + \dot{\theta}^2\left(-\cos\theta - \tan\varphi\sin\theta - \overline{R}\frac{\cos^2\theta}{\cos^3\varphi}\right)\right]x_0 \qquad [1.16]$$

When the crank shaft turns at a constant velocity ω (no acyclism), these equations can be reduced to:

$$\ddot{\varphi} = \overline{R}\dot{\theta}^2\frac{\cos^2\varphi\sin\theta + \overline{R}\cos^2\theta\sin\varphi}{\cos^3\varphi} \qquad [1.17]$$

$$\Gamma(G_2/\Re_0) = R\dot{\theta}^2\left[\left(-\cos\theta - k\tan\varphi\sin\theta - k\overline{R}\frac{\cos^2\theta}{\cos^3\varphi}\right)x_0 - (1-k)\sin\theta y_0\right] \qquad [1.18]$$

$$\Gamma(B/\Re_0) = \ddot{x}x_0 = R\dot{\theta}^2\left(-\cos\theta - \tan\varphi\sin\theta - \overline{R}\frac{\cos^2\theta}{\cos^3\varphi}\right)x_0 \qquad [1.19]$$

1.2. Efforts in the links between the crank shaft, the connecting rod and the piston

1.2.1. *Hypothesis and data*

The single-cylinder model defined in section 1.1 is considered. It will be assumed that the input parameter from the point of view of forces consists of pressure exerted by gas on the upper face of the piston. This pressure varies continuously throughout the whole engine cycle. The piston movement, which results from the aforementioned pressure, is transmitted throughout the entire moving part onto the receiving unit. Because of its large inertia, the latter is assumed to turn at a constant speed, making equations [1.17]–[1.19] valid.

The weight of the elements is assumed to be negligible when compared to the other forces acting on various elements of the mechanism.

The calculations of frictional forces between the piston and the cylinder, either at piston ring or piston skirt level, require an understanding of hydrodynamic behavior of lubricated contacts [GAM 06] that goes beyond the scope of this book.

Tian *et al.* [TIA 96] showed that for a device of comparable dimensions, the frictional forces at skirt/liner contact points do not exceed 50 N, except for several degrees around the top dead center (TDC) region, where end compression occurs, and it reaches 100 N. As far as the frictional forces between the liner and piston are concerned, Keribar and Dursunkaya [KER 92] found that during the combustion phase for a piston with a diameter of 125 mm (50% larger than the one considered here), the hydrodynamic forces were about 200 N and the forces resulting from a mixed lubrication model were around 700 N. Similar values were obtained by Mansouri and Wong [MAN 05]. In the following discussion, these frictional forces are not taken into consideration. The validity of the hypothesis will be checked.

The reference frame $\mathfrak{R}_0$ is assumed to be Galilean. The engine block acceleration due to the movement of the vehicle and itself with respect to the latter is not taken into account.

For the studies of bearing hydrodynamic properties, general dynamic equations are used to determine the forces acting in each link. These forces are divided into x_0 and y_0 components. The z_0 component of the resulting forces is not relevant to the study of bearings, which is why in the following description it will not be considered. The x_0 and y_0 component moment projections that originate from the pressure field in the film are zero for aligned bearings and give the righting moments for non-aligned bearings. As the junctions are assumed to have perfect kinematic properties, this last case is not considered. The projection z_0 component of the moment due to shear stress because of lubricant viscosity is unknown.

Only the equations used for determining the link forces are given.

The given mass parameters are as follows:

– m_1: crank shaft mass;

– m_2: connecting rod mass;

– m_3: piston pin mass;

– m_4: piston mass;

– I_2: moment of inertia of the rod with respect to the axis G z_0;

– I_3: moment of inertia of the axis with respect to the axis B z_0.

1.2.2. *Dynamics equations for the piston*

External forces

– $\mathbf{F}_g = -p_g S \mathbf{x}_0$: force of gas where p_g is the gas pressure and S is the piston surface;

– $\mathbf{F}_{0\to 4} = Y_{04}\mathbf{y}_0$: force exerted by the cylinder on the piston normal to the direction of the displacement of the piston;

– $\mathbf{F}_{3\to 4} = -X_{43}\mathbf{x}_0 - Y_{43}\mathbf{y}_0$: force exerted by the axis on the piston.

Equations

– $m_4 \Gamma(B/\mathfrak{R}_0) = \mathbf{F}_g + \mathbf{F}_{0\to 4} + \mathbf{F}_{3\to 4}$

which projected on basis B_0 gives:

$$\begin{cases} m_4 \ddot{x} = -p_g S - X_{43} \\ 0 = Y_{04} - Y_{43} \end{cases} \qquad [1.20]$$

1.2.3. *Dynamics equations for the axis*

External forces and moments

– $\mathbf{F}_{4\to 3} = X_{43}\mathbf{x}_0 + Y_{43}\mathbf{y}_0$: force exerted by the piston on the axis;

- $\mathbf{F}_{2\to 3} = X_{23}\mathbf{x}_0 + Y_{23}\mathbf{y}_0$: force exerted by the rod on the axis;
- $\mathbf{M}_{4\to 3}(B).\mathbf{z}_0 = M_{43}$: frictional moment exerted by the piston on the axis;
- $\mathbf{M}_{2\to 3}(B).\mathbf{z}_0 = M_{23}$: frictional moment exerted by the rod on the axis.

Equations

- $m_3 \Gamma(B/\mathfrak{R}_0) = \mathbf{F}_{4\to 3} + \mathbf{F}_{2\to 3}$

- $I_3 \ddot{\chi} \, \mathbf{z}_0 = \mathbf{M}_{4\to 3}(B) + \mathbf{M}_{2\to 3}(B)$

which can be combined as follows:

$$\begin{cases} m_3 \ddot{x} = X_{43} + X_{23} \\ 0 = Y_{43} + Y_{23} \\ I_3 \ddot{\chi} = M_{43} + M_{23} \end{cases} \qquad [1.21]$$

1.2.4. Dynamics equations for the connecting rod

External forces and moments

- $\mathbf{F}_{3\to 2} = -X_{23}\mathbf{x}_0 - Y_{23}\mathbf{y}_0$: force exerted by the axis on the connecting rod;
- $\mathbf{F}_{1\to 2} = X_{12}\mathbf{x}_0 + Y_{12}\mathbf{y}_0$: force exerted by the crank shaft on the connecting rod;
- $\mathbf{M}_{3\to 2}(B).\mathbf{z}_0 = -M_{23}$: frictional moment exerted by the axis on the connecting rod;
- $\mathbf{M}_{1\to 2}(A).\mathbf{z}_0 = M_{12}$: frictional moment exerted by the crank shaft on the connecting rod.

Equations

- $m_2 \Gamma(G_2/\mathfrak{R}_0) = \mathbf{F}_{3\to 2} + \mathbf{F}_{1\to 2}$

- $I_2 \ddot{\varphi} \, \mathbf{z}_0 = \mathbf{M}_{3\to 2}(B) + \mathbf{G_2B} \wedge \mathbf{F}_{3\to 2} + \mathbf{M}_{1\to 2}(A) + \mathbf{G_2A} \wedge \mathbf{F}_{1\to 2}$

which can be combined to give:

$$\begin{cases} m_2 R \dot{\theta}^2 \left(-\cos\theta - k\tan\varphi \sin\theta - k\overline{R}\dfrac{\cos^2\theta}{\cos^3\varphi} \right) = -X_{23} + X_{12} \\ -m_2(1-k)R\dot{\theta}^2 \sin\theta = -Y_{23} + Y_{12} \qquad\qquad [1.22] \\ I_2 \ddot{\varphi} = -M_{23} + M_{12} - L\left[(1-k)Y_{23} + kY_{12}\right]\cos\varphi + L\left[(1-k)X_{23} + kX_{12}\right]\sin\varphi \end{cases}$$

There are seven junction strength components and three moment components in equations [1.20]–[1.22]. The unknowns related to angular acceleration of axis $\ddot{\chi}$ should be added to these unknowns, giving a total of 11 unknowns in eight equations. The moments M_{12}, M_{23} and M_{43} related to viscous friction can only be determined by carrying out hydrodynamic calculations.

To resolve this issue, the first possibility is to assume that the viscous friction has a negligible effect on junction force component calculations. Assuming M_{12}, M_{23} and M_{43} are zero, the angular acceleration of axis $\ddot{\chi}$ should also be zero. This leaves seven unknowns in seven equations.

If it is assumed that the moments of viscous friction are not negligible, a more complicated hydrodynamic model must be created in order to incorporate equations [1.20]–[1.22] (creation of a model like this is outlined in section 3.4).

1.2.5. *Dynamics equations for the crank shaft*

External forces

- $\mathbf{F}_{0\to 1} = -X_{10}\mathbf{x}_0 - Y_{10}\mathbf{y}_0$: force exerted by the engine block on the crank shaft;
- $\mathbf{F}_{2\to 1} = -X_{12}\mathbf{x}_0 - Y_{12}\mathbf{y}_0$: force exerted by the connecting rod on the crank shaft.

Equations

- $m_1 \mathbf{\Gamma}(G_1/\mathfrak{R}_0) = \mathbf{F}_{0\to 1} + \mathbf{F}_{2\to 1}$

which can be expressed as:

$$\begin{cases} -m_1 a \dot{\theta}^2 \cos\theta = -X_{10} - X_{12} \\ -m_1 a \dot{\theta}^2 \sin\theta = -Y_{10} - Y_{12} \end{cases} \quad [1.23]$$

If it is assumed that the crank shaft is statically balanced, this would mean that the center of mass is located on the O z_0 axis and the equations become:

$$\begin{cases} 0 = -X_{10} - X_{12} \\ 0 = -Y_{10} - Y_{12} \end{cases} \quad [1.24]$$

1.2.6. *Efforts for frictionless links*

If the viscous friction is ignored, the system can be reduced to the following five equations:

$$\begin{cases} X_{43} = -m_4 \ddot{x} - p_g S \\ X_{43} + X_{23} = m_3 \ddot{x} \\ X_{23} - X_{12} = m_2 R \dot{\theta}^2 K \\ Y_{23} - Y_{12} = m_2 (1-k) R \dot{\theta}^2 \sin\theta \\ [(1-k)X_{23} + kX_{12}]\sin\varphi - [(1-k)Y_{23} + kY_{12}]\cos\varphi = \dfrac{I_2}{L}\ddot{\varphi} \end{cases}$$

where the factor K is defined as:

$$K = \cos\theta + k\tan\varphi \sin\theta + k\overline{R}\dfrac{\cos^2\theta}{\cos^3\varphi} \quad [1.25]$$

If m_{34} represents the sum of the piston mass and pin mass:

$$m_{34} = m_3 + m_4 \quad [1.26]$$

the solution to the system is

$$X_{43} = -m_4\ddot{x} - p_g S$$

$$X_{23} = m_{34}\ddot{x} + p_g S$$

$$X_{12} = -m_2 R\dot\theta^2 K + m_{34}\ddot{x} + p_g S$$

$$Y_{23} = km_2 R\dot\theta^2 \left[(1-k)\sin\theta - K\tan\varphi\right] + \left(m_{34}\ddot{x} + p_g S\right)\tan\varphi - \frac{I_2}{L\cos\varphi}\ddot\varphi$$

$$Y_{12} = -m_2 R\dot\theta^2 \left[(1-k)^2 \sin\theta + kK\tan\varphi\right] + \left(m_{34}\ddot{x} + p_g S\right)\tan\varphi - \frac{I_2}{L\cos\varphi}\ddot\varphi$$

The forces of interaction between the elements of the moving part, expressed using B_0, are thus:

$$F_{1\to 0}\begin{cases} m_2 R\dot\theta^2 K - m_{34}\ddot{x} - p_g S \\ m_2 R\dot\theta^2 \left[(1-k)^2\sin\theta + kK\tan\varphi\right] - \left(m_{34}\ddot{x} + p_g S\right)\tan\varphi + \frac{I_2}{L\cos\varphi}\ddot\varphi \end{cases} \qquad [1.27]$$

$$F_{1\to 2}\begin{cases} -m_2 R\dot\theta^2 K + m_{34}\ddot{x} + p_g S \\ -m_2 R\dot\theta^2 \left[(1-k)^2\sin\theta + kK\tan\varphi\right] + \left(m_{34}\ddot{x} + p_g S\right)\tan\varphi - \frac{I_2}{L\cos\varphi}\ddot\varphi \end{cases} \qquad [1.28]$$

$$F_{2\to 3}\begin{cases} m_{34}\ddot{x} + p_g S \\ km_2 R\dot\theta^2 \left[(1-k)\sin\theta - K\tan\varphi\right] + \left(m_{34}\ddot{x} + p_g S\right)\tan\varphi - \frac{I_2}{L\cos\varphi}\ddot\varphi \end{cases} \qquad [1.29]$$

$$F_{4\to 3}\begin{cases} -m_4\ddot{x} - p_g S \\ -km_2 R\dot\theta^2 \left[(1-k)\sin\theta - K\tan\varphi\right] - \left(m_{34}\ddot{x} + p_g S\right)\tan\varphi + \frac{I_2}{L\cos\varphi}\ddot\varphi \end{cases} \qquad [1.30]$$

1.3. Load diagram correction in the case of large deformations

In the previous sections, the relationship of the force exerted by the crank shaft on the connecting rod is determined based on the pressure of gas in a combustion chamber with a configuration that uses tight junctions with non-deformable solids. For high power engines (for example, Formula 1 engines), the connecting rods are as small as possible to reduce their force dynamic effects. The dynamic load on the bearings still remains high and when passing through the TDC, the traction on the connecting rod is such that the clearance in the bearing exceeds its nominal value several times. In this case, the load diagram can be adjusted as discussed in the following.

1.3.1. *Kinematics of crank shaft–connecting rod–piston system with mobility*

Figure 1.2 shows the crank shaft–connecting rod system when it has mobility in the "small end". The connecting rod end A at the bearing housing center is located at a different position than the end A_1 at the center of the pin. The position of A_1 in the reference frame A $x_2y_2z_2$ in the connecting rod is given using ε_x and ε_y.

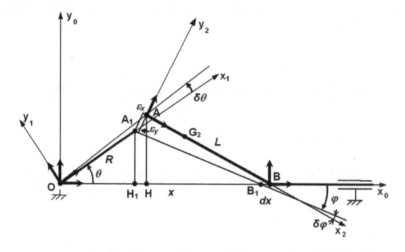

Figure 1.2. *Crank shaft–connecting rod–piston system with mobility in the connecting rod big end link*

By using the previously established notations, the geometric relationships for H_1A_1 and HA can be written as follows:

$$R\sin\theta = -L\sin\varphi$$

$$R\sin\theta - L\sin\varphi - \varepsilon_x \sin\tilde{\varphi} - \varepsilon_y \cos\tilde{\varphi} = -L\sin\tilde{\varphi}$$

with:

$$\tilde{\varphi} = \varphi + \delta\varphi \qquad [1.31]$$

where $\delta\varphi$ is the modification of φ due to mobility. Assuming the displacement of A_1A is small with respect to R, the equations can be written as a function of ε_x and ε_y:

$$\delta\varphi \approx \frac{\varepsilon_y \cos\varphi + \varepsilon_x \sin\varphi}{\varepsilon_y \sin\varphi - (L-\varepsilon_x)\cos\varphi} \approx -\frac{1}{L}\left(\varepsilon_y + \varepsilon_x \tan\varphi\right) \qquad [1.32]$$

The position and speed of the center A of the connecting rod "small end" bearing can be written as:

$$OA = Rx_1 - \varepsilon_x x_2 - \varepsilon_y y_2$$

$$V_{02}(A) = R\omega y_1 - \dot\varepsilon_x x_2 - \dot\varepsilon_y y_2 - \varepsilon_x \dot x_2 - \varepsilon_y \dot y_2 \qquad [1.33]$$

The vectors y_1, x_2 and y_2 can easily be expressed in terms of their components in the basis B_0 and are given as:

$$y_1 = -\sin\theta x_0 + \cos\theta y_0$$

$$x_2 = \cos\tilde\varphi\, x_0 + \sin\tilde\varphi\, y_0 \approx (\cos\varphi - \delta\varphi\sin\varphi)x_0 + (\sin\varphi + \delta\varphi\cos\varphi)y_0$$

$$y_2 = -\sin\tilde\varphi\, x_0 + \cos\tilde\varphi\, y_0 \approx -(\sin\varphi + \delta\varphi\cos\varphi)x_0 + (\cos\varphi - \delta\varphi\sin\varphi)y_0$$

The derivatives of vectors x_2 and y_2 are given as:

$$\dot x_2 = \dot{\tilde\varphi}\left[-\sin\tilde\varphi\, x_0 + \cos\tilde\varphi\, y_0\right] = \left(\dot\varphi + \dot{\delta\varphi}\right)\left[-\sin\tilde\varphi\, x_0 + \cos\tilde\varphi\, y_0\right]$$

$$\dot y_2 = -\dot{\tilde\varphi}\left[\cos\tilde\varphi\, x_0 + \sin\tilde\varphi\, y_0\right] = -\left(\dot\varphi + \dot{\delta\varphi}\right)\left[\cos\tilde\varphi\, x_0 + \sin\tilde\varphi\, y_0\right]$$

with:

$$\dot\varphi = -\overline{R}\omega \frac{\cos\theta}{\cos\varphi} \qquad [1.34]$$

$$\dot{\delta\varphi} = -\frac{1}{L}\left(\dot{\varepsilon}_y + \dot{\varepsilon}_x \tan\varphi + \varepsilon_x \frac{\dot{\varphi}}{\cos^2\varphi}\right) \qquad [1.35]$$

These relationships allow $V_{02}(A)$ to be expressed in the following form:

$$V_{02}(A) = u_A x_0 + v_A y_0 \qquad [1.36]$$

where u_A and v_A can be obtained by substituting the previous equations in [1.33]:

$$\begin{aligned}u_A = -R\omega\sin\theta &+ \left(\dot{\varepsilon}_y + \dot{\varepsilon}_x\,\delta\varphi\right)\sin\varphi - \left(\dot{\varepsilon}_x - \dot{\varepsilon}_y\,\delta\varphi\right)\cos\varphi \\ &+ \dot{\tilde{\varphi}}\left[\left(\varepsilon_x - \varepsilon_y\delta\varphi\right)\sin\varphi + \left(\varepsilon_y + \varepsilon_x\delta\varphi\right)\cos\varphi\right]\end{aligned} \qquad [1.37]$$

$$\begin{aligned}v_A = R\omega(\cos\theta) &- \left(\dot{\varepsilon}_x - \dot{\varepsilon}_y\,\delta\varphi\right)\sin\varphi - \left(\dot{\varepsilon}_y + \dot{\varepsilon}_x\,\delta\varphi\right)\cos\varphi \\ &+ \dot{\tilde{\varphi}}\left[\left(\varepsilon_y + \varepsilon_x\delta\varphi\right)\sin\varphi - \left(\varepsilon_x - \varepsilon_y\delta\varphi\right)\cos\varphi\right]\end{aligned} \qquad [1.38]$$

The velocity of the connecting rod mass center G_2 can be derived as:

$$V_{02}(G_2) = V_{02}(A) + \Omega_{02} \wedge AG_2 = V_{02}(A) + \dot{\tilde{\varphi}} z_0 \wedge kLx_2$$
$$= u_A x_0 + v_A y_0 + kL\dot{\tilde{\varphi}} y_2$$

or:

$$V_{02}(G_2) = \left[u_A - kL\dot{\tilde{\varphi}}(\sin\varphi + \delta\varphi\cos\varphi)\right]x_0 + \left[v_A + kL\dot{\tilde{\varphi}}(\cos\varphi - \delta\varphi\sin\varphi)\right]y_0 \quad [1.39]$$

When the connection at A has no mobility (ε_x, ε_y and $\delta\varphi$ are zero as well as their derivatives), the velocity of G_2 can be simply written as:

$$V_{02}(G_2) = R\omega\left[(-\sin\theta + k\tan\varphi\cos\theta)x_0 + (1-k)\cos\theta y_0\right]$$

The velocity of point B can be calculated by choosing k equal to 1 in the following expression:

$$V_{02}(B) = \left[u_A - L\tilde{\dot{\varphi}}(\sin\varphi + \delta\varphi\cos\varphi) \right] x_0 \qquad [1.40]$$

After deriving [1.39] and [1.40] and eliminating the second-order parameters with respect to $\delta\varphi$, the equations describing acceleration for points G_2 and B can be obtained as:

$$\Gamma(G_2/\Re_0) = \left[\dot{u}_A - kL\left(\left(\ddot{\varphi} + \ddot{\delta\varphi} - \dot{\varphi}^2\,\delta\varphi\right)\sin\varphi + \left(\ddot{\varphi}\,\delta\varphi + \dot{\varphi}^2 + 2\dot{\varphi}\dot{\delta\varphi}\right)\cos\varphi \right) \right] x_0$$
$$+ \left[\dot{v}_A + kL\left(\left(\ddot{\varphi} + \ddot{\delta\varphi} - \dot{\varphi}^2\,\delta\varphi\right)\cos\varphi - \left(\ddot{\varphi}\,\delta\varphi + \dot{\varphi}^2 + 2\dot{\varphi}\dot{\delta\varphi}\right)\sin\varphi \right) \right] y_0 \qquad [1.41]$$

$$\Gamma(B_2/\Re_0) = \left[\dot{u}_A - L\left(\left(\ddot{\varphi} + \ddot{\delta\varphi} - \dot{\varphi}^2\,\delta\varphi\right)\sin\varphi + \left(\ddot{\varphi}\,\delta\varphi + \dot{\varphi}^2 + 2\dot{\varphi}\dot{\delta\varphi}\right)\cos\varphi \right) \right] x_0 \qquad [1.42]$$

$\ddot{\tilde{\varphi}}$ is obtained by deriving $\dot{\varphi} + \ddot{\delta\varphi}$ (equations [1.34] and [1.35]) assuming the crank shaft turns at a constant speed ω (no acyclism):

$$\ddot{\varphi} = \overline{R}\omega\,\frac{\omega\cos\varphi\sin\theta - \dot{\varphi}\sin\varphi\cos\theta}{\cos^2\varphi}$$

$$\ddot{\delta\varphi} = -\frac{1}{L}\left[\ddot{\varepsilon}_y + \ddot{\varepsilon}_x\tan\varphi + \frac{1}{\cos^2\varphi}\left(\ddot{\varepsilon}_x\,\dot{\varphi} + 2\dot{\varepsilon}_x\,\dot{\varphi}^2\tan\varphi + 2\dot{\varepsilon}_x\,\ddot{\varphi} \right) \right] \qquad [1.43]$$

The parameters $\dot{u}_A$ and $\dot{v}_A$ can be derived by using equations [1.37] and [1.38] and eliminating the second-order parameters with respect to $\delta\varphi$ as follows:

$$\begin{aligned}\dot{u}_A = & -R\omega^2 \cos\theta + \left[\left(\varepsilon_x - \varepsilon_y \delta\varphi\right)\cos\varphi - \left(\varepsilon_y + \varepsilon_x \delta\varphi\right)\sin\varphi\right]\dot{\varphi}^2 \\ & + \left[\ddot{\varepsilon}_y + \ddot{\varepsilon}_x \delta\varphi + 2\left(\dot{\varepsilon}_x - \dot{\varepsilon}_y \delta\varphi\right)\dot{\varphi} + 2\left(\dot{\varepsilon}_x - \varepsilon_y \dot{\varphi}\right)\dot{\delta\varphi}\right]\sin\varphi \\ & - \left[\ddot{\varepsilon}_x - \ddot{\varepsilon}_y \delta\varphi - 2\left(\dot{\varepsilon}_y + \dot{\varepsilon}_x \delta\varphi\right)\dot{\varphi} - 2\left(\dot{\varepsilon}_y + \varepsilon_x \dot{\varphi}\right)\dot{\delta\varphi}\right]\cos\varphi \\ & + \ddot{\varphi}\left[\left(\varepsilon_x - \varepsilon_y \delta\varphi\right)\sin\varphi + \left(\varepsilon_y + \varepsilon_x \delta\varphi\right)\cos\varphi\right] + \ddot{\delta\varphi}\left(\varepsilon_x \sin\varphi + \varepsilon_y \cos\varphi\right)\end{aligned}$$

[1.44]

$$\begin{aligned}\dot{v}_A = & -R\omega^2 \sin\theta + \left[\left(\varepsilon_y + \varepsilon_x \delta\varphi\right)\cos\varphi + \left(\varepsilon_x - \varepsilon_y \delta\varphi\right)\sin\varphi\right]\dot{\varphi}^2 \\ & - \left[\ddot{\varepsilon}_x - \ddot{\varepsilon}_y \delta\varphi - 2\left(\dot{\varepsilon}_y + \dot{\varepsilon}_x \delta\varphi\right)\dot{\varphi} - 2\left(\dot{\varepsilon}_y + \varepsilon_x \dot{\varphi}\right)\dot{\delta\varphi}\right]\sin\varphi \\ & - \left[\ddot{\varepsilon}_y + \ddot{\varepsilon}_x \delta\varphi + 2\left(\dot{\varepsilon}_x - \dot{\varepsilon}_y \delta\varphi\right)\dot{\varphi} + 2\left(\dot{\varepsilon}_x - \varepsilon_y \dot{\varphi}\right)\dot{\delta\varphi}\right]\cos\varphi \\ & + \ddot{\varphi}\left[\left(\varepsilon_y + \varepsilon_x \delta\varphi\right)\sin\varphi - \left(\varepsilon_x - \varepsilon_y \delta\varphi\right)\cos\varphi\right] + \ddot{\delta\varphi}\left[\varepsilon_y \sin\varphi - \varepsilon_x \cos\varphi\right]\end{aligned}$$

[1.45]

For simplification reasons, the following notation is defined:

$$\Gamma(G_2/\mathfrak{R}_0) = \left(\gamma_x^{G_2} + \delta\gamma_x^{G_2}\right)x_0 + \left(\gamma_y^{G_2} + \delta\gamma_y^{G_2}\right)y_0$$

[1.46]

where $\gamma_x^{G_2}$ and $\gamma_y^{G_2}$ represent the components of acceleration of B in the case of a connection at A without mobility (section 1.1.4):

$$\begin{cases}\gamma_x^{G_2} = -R\omega^2\left(\cos\theta + k\tan\varphi\sin\theta + k\bar{R}\dfrac{\cos^2\theta}{\cos^3\varphi}\right) \\ \gamma_y^{G_2} = -R\omega^2(1-k)\sin\theta\end{cases}$$

[1.47]

From these relationships and equations [1.41], the following equations can be deduced:

$$\delta\gamma_x^{G_2} = \left[-(\varepsilon_y \cos\varphi + \varepsilon_x \sin\varphi)\dot{\varphi}^2 + (\varepsilon_x \cos\varphi - \varepsilon_y \sin\varphi)\ddot{\varphi} \right]\delta\varphi$$

$$+ \left[\left(\ddot{\varepsilon}_x - 2\dot{\varepsilon}_y \dot{\varphi} \right)\sin\varphi + \left(\ddot{\varepsilon}_y + 2\dot{\varepsilon}_x \dot{\varphi} \right)\cos\varphi + kL\left(\dot{\varphi}^2 \sin\varphi - \ddot{\varphi}\cos\varphi \right) \right]\delta\varphi \quad [1.48]$$

$$+ 2\left[\left(\dot{\varepsilon}_y + \varepsilon_x \dot{\varphi} \right)\cos\varphi + \left(\dot{\varepsilon}_x - \varepsilon_y \dot{\varphi} \right)\sin\varphi - kL\dot{\varphi}\cos\varphi \right]\dot{\delta\varphi}$$

$$+ (\varepsilon_x \sin\varphi + \varepsilon_y \cos\varphi - kL\sin\varphi)\ddot{\delta\varphi}$$

$$\delta\gamma_y^{G_2} = \left[(\varepsilon_x \cos\varphi - \varepsilon_y \sin\varphi)\dot{\varphi}^2 + (\varepsilon_x \sin\varphi + \varepsilon_y \cos\varphi)\ddot{\varphi} \right]\delta\varphi$$

$$+ \left[\left(\ddot{\varepsilon}_y + 2\dot{\varepsilon}_x \dot{\varphi} \right)\sin\varphi - \left(\ddot{\varepsilon}_x - 2\dot{\varepsilon}_y \dot{\varphi} \right)\cos\varphi - kL\left(\dot{\varphi}^2 \cos\varphi + \ddot{\varphi}\sin\varphi \right) \right]\delta\varphi \quad [1.49]$$

$$+ 2\left[\left(\dot{\varepsilon}_y + \varepsilon_x \dot{\varphi} \right)\sin\varphi - \left(\dot{\varepsilon}_x - \varepsilon_y \dot{\varphi} \right)\cos\varphi - kL\dot{\varphi}\sin\varphi \right]\dot{\delta\varphi}$$

$$+ (\varepsilon_y \sin\varphi - \varepsilon_x \cos\varphi + kL\cos\varphi)\ddot{\delta\varphi}$$

$$V_{02}(B) = \left[u_A - L\dot{\tilde{\varphi}}(\sin\varphi + \delta\varphi\cos\varphi) \right]x_0$$

Similarly, it can be written that:

$$\Gamma(B/\mathfrak{R}_0) = \left(\ddot{x} + \ddot{\delta x} \right)x_0 \quad [1.50]$$

where $\ddot{x}$ represents the component of acceleration of B in the case of a connection at A without mobility (section 1.1.4):

$$\ddot{x} = -R\omega^2 \left(\cos\theta + \tan\varphi \sin\theta + \bar{R}\frac{\cos^2\theta}{\cos^3\varphi} \right)$$

$\ddot{\delta x}$ can be obtained using $\gamma_x^{G_2}$ (equation [1.19]) assuming that k is equal to 1.

1.3.2. Dynamics of crank shaft–connecting rod–piston system with mobility

For junction force component calculations, using the mass data defined in section 1.2.1 and assuming that viscous friction is negligible, the following equations can be derived by the laws of dynamics:

– Dynamic equations for the piston

$$\begin{cases} m_4\left(\ddot{x}+\ddot{\delta x}\right) = -p_g S - X_{43} \\ 0 = Y_{04} - Y_{43} \end{cases}$$

– Dynamic equations for the axis

$$\begin{cases} m_3 \ddot{x} = X_{43} + X_{23} \\ 0 = Y_{43} + Y_{23} \end{cases}$$

– Dynamic equations for the connecting rod

$$\begin{cases} m_2\left(\gamma_x^{G_2} + \delta\gamma_x^{G_2}\right) = -X_{23} + X_{12} \\ m_2\left(\gamma_y^{G_2} + \delta\gamma_y^{G_2}\right) = -Y_{23} + Y_{12} \\ I_2 \ddot{\tilde{\varphi}} = -L[(1-k)Y_{23} + kY_{12}]\cos\tilde{\varphi} + L[(1-k)X_{23} + kX_{12}]\sin\tilde{\varphi} \end{cases}$$

The system can be reduced to the following five equations:

$$\begin{cases} X_{43} = -m_4\left(\ddot{x}+\ddot{\delta x}\right) - p_g S \\ X_{43} + X_{23} = m_3\left(\ddot{x}+\ddot{\delta x}\right) \\ X_{23} - X_{12} = -m_2\left(\gamma_x^{G_2} + \delta\gamma_x^{G_2}\right) \\ Y_{23} - Y_{12} = -m_2\left(\gamma_y^{G_2} + \delta\gamma_y^{G_2}\right) \\ [(1-k)X_{23} + kX_{12}]\sin\tilde{\varphi} - [(1-k)Y_{23} + kY_{12}]\cos\tilde{\varphi} = \dfrac{I_2}{L}\ddot{\tilde{\varphi}} \end{cases}$$

If m_{34} represents the sum of the masses of the piston and the piston pin:

$$m_{34} = m_3 + m_4$$

the system solution is given as:

$$X_{43} = -m_4\left(\ddot{x} + \ddot{\delta x}\right) - p_g S$$

$$X_{23} = m_{34}\left(\ddot{x} + \ddot{\delta x}\right) + p_g S$$

$$Y_{23} = -km_2\left(\gamma_y^{G_2} + \delta\gamma_y^{G_2}\right) + \left[m_{34}\left(\ddot{x} + \ddot{\delta x}\right) + p_g S + km_2\left(\gamma_x^{G_2} + \delta\gamma_x^{G_2}\right)\right]\tan\tilde{\varphi} - \frac{l_2}{L}\frac{\ddot{\tilde{\varphi}}}{\cos\tilde{\varphi}}$$

$$X_{12} = m_2\left(\gamma_x^{G_2} + \delta\gamma_x^{G_2}\right) + m_{34}\left(\ddot{x} + \ddot{\delta x}\right) + p_g S$$

$$Y_{12} = (1-k)m_2\left(\gamma_y^{G_2} + \delta\gamma_y^{G_2}\right) + \left[m_{34}\left(\ddot{x} + \ddot{\delta x}\right) + p_g S + km_2\left(\gamma_x^{G_2} + \delta\gamma_x^{G_2}\right)\right]\tan\tilde{\varphi} - \frac{l_2}{L}\frac{\ddot{\tilde{\varphi}}}{\cos\tilde{\varphi}}$$

For TEHD bearing studies, the force $\mathbf{F}_{1\to 2}$ is expressed in the basis B_2 attached to the connecting rod as:

$$\mathbf{F}_{1\to 2} = X_{12}\mathbf{x}_0 + Y_{12}\mathbf{y}_0$$
$$= \left[X_{12}\cos\tilde{\varphi} + Y_{12}\sin\tilde{\varphi}\right]\mathbf{x}_2 + \left[Y_{12}\cos\tilde{\varphi} - X_{12}\sin\tilde{\varphi}\right]\mathbf{y}_2$$

which gives:

$$\mathbf{F}_{1\to 2} = \left[(1-k)m_2\left(\left(\gamma_x^{G_2} + \delta\gamma_x^{G_2}\right)\cos\tilde{\varphi} + \left(\gamma_y^{G_2} + \delta\gamma_y^{G_2}\right)\sin\tilde{\varphi}\right)\right.$$
$$+\left.\left(km_2\left(\gamma_x^{G_2} + \delta\gamma_x^{G_2}\right) + m_{34}\left(\ddot{x} + \ddot{\delta x}\right) + p_g S - \frac{l_2}{L}\ddot{\tilde{\varphi}}\sin\tilde{\varphi}\right)\frac{1}{\cos\tilde{\varphi}}\right]\mathbf{x}_2 \quad [1.51]$$
$$+\left[(1-k)m_2\left(\left(\gamma_y^{G_2} + \delta\gamma_y^{G_2}\right)\cos\tilde{\varphi} - \left(\gamma_x^{G_2} + \delta\gamma_x^{G_2}\right)\sin\tilde{\varphi}\right) - \frac{l_2}{L}\ddot{\tilde{\varphi}}\right]\mathbf{y}_2$$

If W_x and W_y represent the components of a calculated load diagram for a without mobility configuration at A, the equilibrium equations describing the pressure field (see section 2.3.1 [BON 14]) must take into consideration the loading change:

$$\iint_\Omega [p(x,z)+p_c(x,z)]\sin\frac{x}{R}dxdz = F_{1\to 2}\cdot x_2 = W_x + \delta F_{1\to 2}\cdot x_2$$

$$\iint_\Omega [p(x,z)+p_c(x,z)]\cos\frac{x}{R}dxdz = \delta F_{1\to 2}\cdot y_2 = W_y + \delta F_{1\to 2}\cdot y_2$$

where $\delta F_{1\to 2}$ represents the correction of the load, which is derived from the connection mobility between the crank shaft and the connecting rod. Considering the approximate first-order equations of $\cos\tilde{\varphi}$ and $\sin\tilde{\varphi}$, $\delta F_{1\to 2}$ can be expressed using the basis B_2 as:

$$\begin{aligned}\delta F_{1\to 2} = &\left[(1-k)m_2\left(\cos\varphi\,\delta\gamma_x^{G_2}+\sin\varphi\,\delta\gamma_y^{G_2}-\left(\gamma_x^{G_2}\sin\varphi-\gamma_y^{G_2}\cos\varphi\right)\delta\varphi\right)\right.\\&+\left(km_2\,\delta\gamma_x^{G_2}+m_{34}\,\ddot{\delta x}-\frac{I_2}{L}\sin\varphi\,\ddot{\delta\varphi}\right)\frac{1}{\cos\varphi}\\&\left.+\left(\left(km_2\gamma_x^{G_2}+m_{34}\,\ddot{x}+p_gS\right)\sin\varphi-\frac{I_2}{L}\ddot{\varphi}\right)\frac{\delta\varphi}{\cos^2\varphi}\right]x_2\qquad [1.52]\\&+\left[(1-k)m_2\left(-\sin\varphi\,\delta\gamma_x^{G_2}+\cos\varphi\,\delta\gamma_y^{G_2}-\left(\gamma_x^{G_2}\cos\varphi+\gamma_y^{G_2}\sin\varphi\right)\delta\varphi\right)-\frac{I_2}{L}\ddot{\delta\varphi}\right]y_2\end{aligned}$$

The load components without mobility can be given by:

$$W_x = (1-k)m_2\left(\gamma_x^{G_2}\cos\varphi+\gamma_y^{G_2}\sin\varphi\right)+\left(km_2\gamma_x^{G_2}+m_{34}\,\ddot{x}+p_gS-\frac{I_2}{L}\ddot{\varphi}\sin\varphi\right)\frac{1}{\cos\varphi}$$

$$W_y = (1-k)m_2\left(-\gamma_x^{G_2}\sin\varphi+\gamma_y^{G_2}\cos\varphi\right)-\frac{I_2}{L}\ddot{\varphi}$$

Using the equation for W_x, it is possible to express $\delta F_{1 \to 2}$ with respect to basis B_2 regardless of the gas pressure as:

$$\delta F_{1\to 2} = \left[(1-k)m_2 \left(\cos\varphi\, \delta\gamma_x^{G_2} + \sin\varphi\, \delta\gamma_y^{G_2} - \left(\gamma_x^{G_2}\sin\varphi - \gamma_y^{G_2}\cos\varphi\right)\delta\varphi \right) \right.$$
$$+ \left(km_2\, \delta\gamma_x^{G_2} + m_{34}\, \ddot{\delta}x - \frac{l_2}{L}\sin\varphi\, \ddot{\delta\varphi} \right) \frac{1}{\cos\varphi}$$
$$+ \left(\left(W_x - (1-k)m_2 \left(\gamma_x^{G_2}\cos\varphi + \gamma_y^{G_2}\sin\varphi \right) \right) \tan\varphi - \frac{l_2}{L}\ddot\varphi \right)\delta\varphi \bigg] x_2 \qquad [1.53]$$
$$+ \left[(1-k)m_2 \left(-\sin\varphi\, \delta\gamma_x^{G_2} + \cos\varphi\, \delta\gamma_y^{G_2} - \left(\gamma_x^{G_2}\cos\varphi + \gamma_y^{G_2}\sin\varphi\right)\delta\varphi \right) - \frac{l_2}{L}\ddot{\delta\varphi} \right] y_2$$

1.4. Examples of link efforts between the elements of crank shaft–connecting rod–piston system

1.4.1. *Data*

A single-cylinder model as defined in section 1.1 is examined with given parameters summed up in Table 1.1.

Figure 1.3 shows a diagram of the pressure applied to the piston surface. This corresponds to a four-stroke engine, with the beginning of the intake phase chosen as the zero engine angle. To demonstrate the effect of inertia forces on bearing load, it is assumed that the diagram of gas pressure is the same in both cases. In reality, differences of more than 20% above the maximum pressure can be measured [SVO 84].

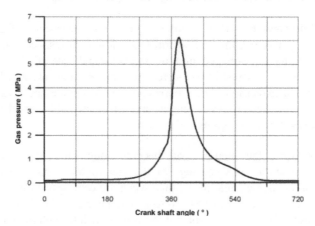

Figure 1.3. *Diagram of gas pressure*

For hydrodynamic calculations, the forces must be projected on the basis relevant to the frame of each bearing housing.

For example, the force $F_{1\to2}$ split into components using the basis B_2 connected to the connecting rod gives:

$$F_{1\to2}/B_2$$

$$\begin{cases} -m_2 R\dot{\theta}^2 \left(K\cos\varphi(1+k\tan^2\varphi)+(1-k)^2\sin\theta\sin\varphi\right)+\dfrac{m_{34}\ddot{x}+p_g S}{\cos\varphi}-\dfrac{I_2}{L}\ddot{\varphi}\tan\varphi \\ (1-k)m_2 R\dot{\theta}^2 \left(K\sin\varphi-(1-k)\sin\theta\cos\varphi\right)-\dfrac{I_2}{L}\ddot{\varphi} \end{cases}$$

To illustrate the different load diagram shapes, the following three motor rotation frequencies are considered: 2,500, 5,000 and 7,500 revolutions per minute (rpm).

Crank shaft radius	R	46.5	mm
Misalignment defect	D	0	mm
Connecting rod length	L	144	mm
Relative position of the con rod mass center	K	0.2628	
Piston surface	S	5,372	mm^2
Connecting rod mass	m_2	0.672	kg
Connecting rod moment of inertia (G z_0)	I_2	2,660	kg·mm^2
Piston and axis mass	$m_3 + m_4$	0.422	kg

Table 1.1. *Data for a monocylinder*

1.4.2. *Load diagrams for the connecting rod big end bearing*

Figure 1.4 shows $F_{1\to2}/B_2$ in a polar diagram for a frequency of 2,500 rpm. At this relatively low speed, the inertial forces are low with respect to the load. They correspond to the olive-shaped part on the left-hand side of the diagram. The connecting rod is essentially subjected to compression produced during gas compression and expansion phases corresponding to engine angle values between 300° and 540° (Figure 1.3). As the engine speed increases, the inertial load increases its amplitude as shown in Figures 1.5 and 1.6.

At 7,500 rpm, the inertial load becomes dominant. The traction on the connecting rod at the TDC passage at the end of the exhaust phase is higher (36, 455 N at an engine angle of 0°) than when the compression is at its maximum (29, 573 N at an engine angle of 511°). At an engine angle of 180°, corresponding to the passage at the bottom dead center at the end of the intake phase, the compression is due solely to inertial forces and reaches 26,557 N.

The force $\mathbf{F}_{1\to 2}$ be expressed in terms of components with respect to the basis B_1 attached to the crank shaft (Figure 1.7). This helps to highlight the position of the force exerted by the connecting rod on the crank shaft pin: it is shown in Figure 1.7 that the force exerted by the connecting rod on the crank pin is never at a force direction angle between 0° and 90°. This area will therefore be the preferred one for lubricant supply placement.

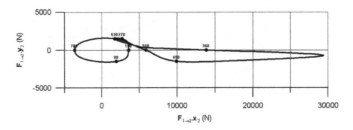

Figure 1.4. *Load polar diagram for a connecting rod big end bearing at 2,500 rpm with respect to the basis attached to the crank shaft*

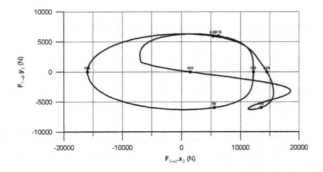

Figure 1.5. *Load polar diagram for a connecting rod big end bearing at 5,000 rpm*

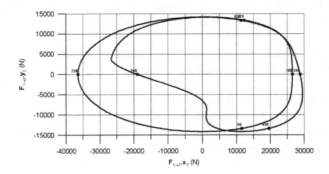

Figure 1.6. *Load polar diagram for a connecting rod big end bearing at 7,500 rpm*

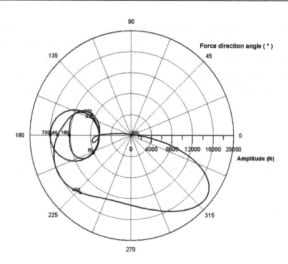

Figure 1.7. *Load polar diagram for a connecting rod big end bearing at 7,500 rpm with respect to the basis attached to the crank shaft*

1.4.3. *Load diagrams for a connecting rod small end bearing*

The polar diagrams showing the force exerted by the connecting rod small end are shown in Figures 1.8–1.10 projected on the basis B_2 attached to the rod for the three regime conditions. At 2,500 rpm, the force is almost exclusively orientated toward the connecting rod.

At 5,000 rpm, the traction on the connecting rod due to the piston mass reaches 6,648 N at the TDC. At 7,500 rpm, the inertial forces are much more pronounced with a traction of 15,544 N at the TDC.

The maximum thrusts on the connecting rod among the three conditions were 31,483, 26,667 and 18,741 N, respectively, at a 383° engine angle, where the pressure is the highest in the combustion chamber (Figure 1.3).

The important difference between these three values well reflects the pressure loss at higher speed due to the inertia of the piston.

The level reached by the forces exerted on the connecting rod small end is about 100 times higher than the magnitude of frictional forces between the cylinder liner on one side and the piston skirt on the other side. Thus, assuming that the mentioned forces are relatively negligible can be justified.

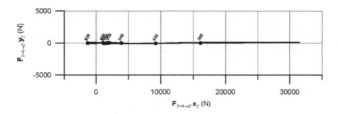

Figure 1.8. *Load polar diagram for a connecting rod small end bearing at 2,500 rpm*

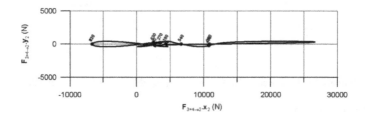

Figure 1.9. *Load polar diagram for a connecting rod small end bearing at 5,000 rpm*

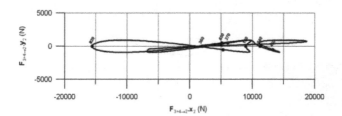

Figure 1.10. *Load polar diagram for a connecting rod small end bearing at 7,500 rpm*

1.4.4. *Load diagrams for a crank shaft main bearing*

The crank shaft of a monocylinder is usually connected to the engine block with two bearings. The diagrams shown in Figures 1.11–1.13 combine the efforts of both bearings in the case of a statistically balanced crank shaft. The pear-shaped part of the curve covering the angular interval 630°–720°/0°–270° corresponds to a completely inertial load at the exhaust and intake phases of the engine cycle. The pressure on the bearing cap marked at 2,500 rpm (29,143 N at an engine angle of 381°) is then gradually absorbed by the inertial forces when increasing the rotational frequency (18,032 N at 5,500 rpm) and becomes negligible compared to the original inertial force at 7,500 rpm.

For this type of high speed, the force acting on the bearing cap is 26,557 N at 180° engine angle. At an engine angle of 0°, the moving part assembly gives rise to a 36,455 N force directed against the engine block.

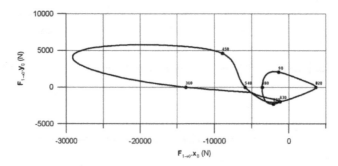

Figure 1.11. *Load polar diagram for a crank shaft main bearing at 2,500 rpm*

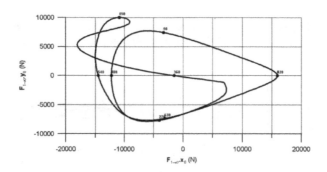

Figure 1.12. *Load polar diagram for a crank shaft main bearing at 5,000 rpm*

1.4.5. Engine torque

Figure 1.14 shows the engine torques for the three regimes. They oscillate between – 530 Nm at 35° engine angle for 7,500 rpm and 853 Nm at 395° engine angle for 2,500 rpm. These significant variations can be caused by the crank shaft speed fluctuations (acyclism), especially if the inertia moment of the wheel is not sufficient enough.

The order of magnitude of the engine torque is more than 100 times greater than the frictional torque in the different bearings of the moving part. Therefore, omitting the latter value in load calculations should not make a major difference.

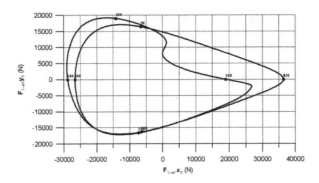

Figure 1.13. *Load polar diagram for a crank shaft main bearing at 7,500 rpm*

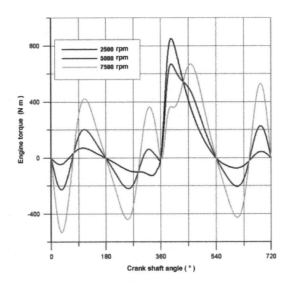

Figure 1.14. *Engine torque for the three rotational speeds*

1.5. Bibliography

[BON 14] BONNEAU D., FATU A., SOUCHET D., *Thermo-hydrodynamic Lubrication in Hydrodynamic Bearings*, ISTE, London and John Wiley & Sons, New York, 2014.

[GAM 06] GAMBIN F., Etude de la consommation d'huile par évaporation dans la chambre de combustion des moteurs thermiques. Développement d'un banc d'essais de prélèvement de gaz, PhD Thesis, University of Poitiers, France, 2006.

[KER 92] KERIBAR R., DURSUNKAYA Z., "A comprehensive model of piston skirt lubrication", SAE Paper 920483, pp. 81–94, 1992.

[MAN 05] MANSOURI S.H., WONG V.W., "Effects of piston design parameters on piston secondary motion and skirt-liner friction", *Journal of Engineering Technology*, vol. 219, pp. 435–449, 2005.

[SVO 84] SVOBODA B., *Mécanique des moteurs alternatifs*, Technip, Paris, 1984.

[TIA 96] TIAN T., WONG V.W., HEYWOODN J.B., "A piston ring-pack film thickness and friction model for multigrade oils and rough surfaces", SAE Paper 962032, pp. 27–39, 1996.

2

The Crank Shaft–Connecting Rod Link

The crank shaft and connecting rod link, which is also referred to as the "connecting rod big end bearing", from a lubrication point of view is definitely one of the best studied parts of the internal combustion engine [BOO 01, BOO 10]. The main reason why this link is of such interest is because it has always been very prone to damage that often results in a fatal outcome for the engine. Until the late 1960s, there were few drivers who had not "thrown out a rod". Since then, progress resulting from numerous experimental and theoretical studies on this link has led to improvements in the shape and materials used for creating crank shafts, connecting rods and their bearings, as well as better lubricant control. Currently, damage related to connecting rod bearings is very rare, at least for regular service vehicles. Before the engine size for a given power can decrease, further studies are needed and numerical modeling has proven to play a significant role here.

2.1. Geometrical and mechanical characteristics of the connecting rod big end bearing

The crank shaft–connecting rod bearing link is the only bearing in the moving part for which the lubrication parameter calculations can be carried out independently of the other bearings. The clearance within the connections of the power train (engine block–crank shaft–connecting rod–piston pin–piston) is sufficiently large to ensure enough mobility to satisfy the non-redundant constraints of the system. This allows for calculations using the *a priori* established load diagram acting on this bearing (see Chapter 1).

In general, the reference frame is centered at the geometric center of the big end of the connecting rod without any load and thus deformation. The axes of this reference frame are those of the connecting rod, for example those defined in Figure 2.1. The crank pin is driven by the main rotational movement around the *Oz* axis (Figure 2.2) together with additional small rigid body movements with amplitudes of ε_x and ε_y and eventually the rotation amplitudes of ζ_x and ζ_y.

Figure 2.1. *Connecting rod big end bearing axis*

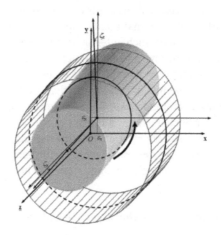

Figure 2.2. *Position parameters of the shaft in the big end housing*

To reduce the inertial forces, the connecting rod has the lightest structure possible. It results in a considerable flexibility that makes it absolutely necessary to take any deformation into consideration when carrying out connecting rod big end bearing lubrication calculations. For engines with a higher rotational frequency (gasoline and competition vehicle engines), the deformations resulting from accelerating fields are not negligible and should therefore also be accounted for.

At the bottom level, the connecting rod bearing is mainly composed of six elastic parts: the connecting rod body, the cap, two shells and screws. Tightening is usually enough to prevent any relative movement at the interfaces between these different bodies. The rod is then considered a single solid with linear elastic behavior. However, under certain conditions of severe loading intensity and/or duration, sliding or separation may occur in some areas of the interfaces. Modeling these phenomena requires specific models presented in section 2.4.

2.2. Lubricant supply

The connecting rod big end bearing is usually supplied by an opening on the crank pin surface. The lubricant reaches the level of this opening by a passage within the crank shaft. This passage is supplied by a circumferential or a semicircumferential groove formed in one of the journal bearings located on either side of the examined crank pin (Figure 2.3). Usually, volumetric gear pumps ensure a flow that evenly covers all the bearings. The pressure at the entrance of each bearing results from the pump outlet pressure and the different pressure drops in the lubrication passages and parts. For the bearings, the flow rate and the pressure loss are closely linked. As the pressure drop due to the thinness of the clearance in the bearings is a lot greater than the losses in the passages, it is considered that the latter is negligible and that the bearing entrance pressure is equal to that of the pump outlet (or the output of a pressure control device if there is one).

For the connecting rod big end bearings and especially for the high power engines, the increase in pressure Δp due to lubricant circulation in the passage, which connects the journal to the crank pin, must be taken into account. This is given by the following equation:

$$\Delta p = \frac{1}{2} \rho_f \omega^2 d^2 \qquad [2.1]$$

where ρ_f is the density of the lubricant, ω is the angular velocity of the crank shaft and d is the distance to the opening from the crank shaft axis (Figure 2.3).

The angular location α of the opening to the crank pin (Figure 2.3) depends on the engine type and the operation conditions: for an engine with a low rotational frequency (diesel engine), the opening is located on the outer part of the crank pin 20–60° before the generatrix furthest from the axis of the crank shaft so as to appropriately replenish the bearings in the phase preceding the end of the compression. For engines with higher frequency (gasoline engines), the opening is located at about 60°.

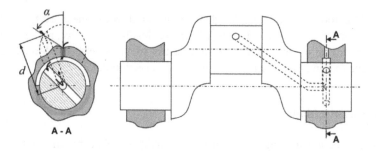

Figure 2.3. *Lubricant supply of a big end connecting rod bearing*

Finite element modeling for the thermo elasto hydro dynamic (TEHD [BON 14a]) problem or more simply for the elastohydro dynamic (EHD, [BON 14b]) problem related to the connecting rod big end bearing uses a mesh attached to the rod. The supply port used on the shaft therefore has a variable position relative to this reference frame. The angle ψ measures the shaft rotation relative to the connecting rod housing (Figure 1.1). The angular position of the opening in the mesh of the film is thus given by the angle $\alpha - \psi$. Aligning the edges of the opening with the boundaries of the mesh requires mesh redefinition at each step. The supply pressure is thus only imposed on nodes that "fall" in the area occupied by the surface opening at that time. The actual opening is elliptical and the digital form of the opening is stepped (see Figure 2.31 [BON 14a]) and includes a variable number of nodes. This lack of coincidence between the digital form and the real shape of the opening is analyzed in detail in section 2.5.3.4, Chapter 2 of volume 3 [BON 14a]. This has no effect on its maximum pressure and temperature values and gives a mean temperature, which can be shifted by a few degrees in a uniform manner over the entire cycle.

2.3. Correction of the load diagram in the case of large deformations

The diagram showing the load exerted by the crank shaft on the connecting rod is determined using the gas pressure diagram in an internal combustion engine without clearance and indeformable solids (Chapter 1). For a high-frequency engine (for example a Formula 1 (F1) engine), the dynamic load exerted on the bearing is very high at the passage point through the top dead center; the traction on the connecting rod is such that the clearance in the bearing exceeds its nominal value several times. In this case, the load diagram can be adjusted as described in Chapter 1.

This section shows the load diagram correction for an F1 engine whose characteristics are outlined in Table 2.1 (only the order of magnitude is given).

Crank shaft radius	R	22	mm
Connecting rod length	L	106	mm
Connecting rod mass	m_2	0.3	kg
Connecting rod moment of inertia	I_2	500	kg·mm^2
Piston mass (including the axis)	m_{34}	0.3	kg
Relative position of the con rod mass center	k	0.4	
Radial clearance (without deformation)	C	20	µm

Table 2.1. *Order of magnitude of data of the linkage bodies for an F1 engine*

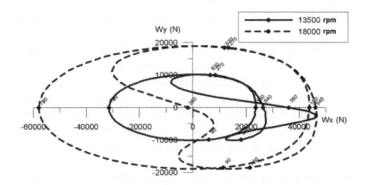

Figure 2.4. *Load diagrams for the con rod big end bearing of an F1 engine*

For this test, the crank pin is regarded as non-deformable. The connecting rod is deformable and displays a linear behavior (no movement between the separate parts). The mentioned deformations result in pressure forces and dynamic effects (see section 4.2.2, Chapter 4 of Volume 1 [BON 14b]). Two engine speeds are considered 13,500 and 18,000 rpm. The load diagrams describing the effect of the crank pin on the big end bearings are shown in Figure 2.4 (with a basis B_2 related to the connecting rod).

The calculations are performed assuming the bearing is isothermal with the same viscosity in both speeds. This hypothesis is not realistic but the purpose of this test is to highlight the differences that may result from using a load correction in the case of a significant movement in the big end bearing. The lubricant displays a piezoviscous nature given by the power law formula:

$$\mu = \mu_0 (1+ap)^b$$

and the coefficient values are given in Table 2.2.

Dynamic viscosity at ambient pressure	μ_0	0.01	Pa·s
Coefficient a of piezoviscosity law	a	0.0036	MPa^{-1}
Coefficient b of piezoviscosity law	b	4.6	

Table 2.2. *Lubricant data*

Figures 2.5 and 2.6 show the orbits around the center of the crank pin for both engine speeds in the first case by considering the load diagrams without load corrections shown in Figure 2.4 and then in the second case with load corrections (see equation [1.53]). The differences between the two cases (with and without load corrections) are very low at 13,500 rpm and slightly more noticeable at 18,000 rpm.

The minimum thickness profiles are slightly modified with the load corrections (Figures 2.7 and 2.8). It is worth noting, however, that *the inclusion of load variations due to eccentricity variations results in a stabilizing effect on the calculation when the change is fast*, even if at the time the load is moderate. This is particularly the case for the 18,000 rpm speed when the engine angle is between 330° and 420°, the phase during which the crank pin "crosses" the bearing housing (Figure 2.6) while the load is near zero at an angle of 360° (Figure 2.5).

Maximum pressure values are only slightly affected by the load corrections (Figures 2.9 and 2.10).

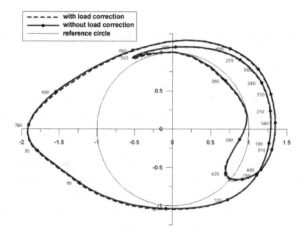

Figure 2.5. *The orbit of a crank pin center at 13,500 rpm*

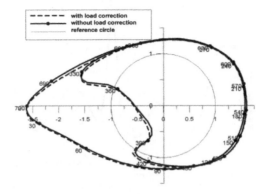

Figure 2.6. *The orbit of a crank pin center at 18,000 rpm*

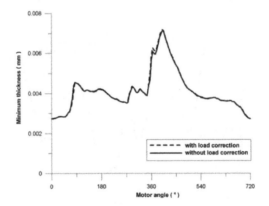

Figure 2.7. *Minimum film thickness at 13,500 rpm*

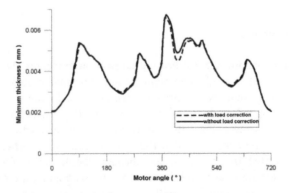

Figure 2.8. *Minimum film thickness at 18,000 rpm*

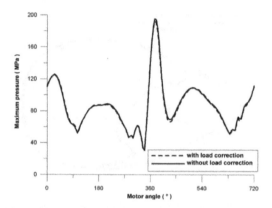

Figure 2.9. *Maximum pressure at 13,500 rpm*

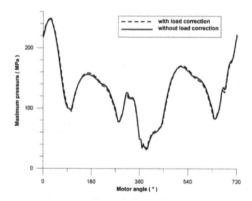

Figure 2.10. *Maximum pressure at 18,000 rpm*

2.4. Multibody models

In elastohydrodynamic big end bearing lubrication modeling, the connecting rod is usually considered a single non-homogenous deformable solid. In reality, the connecting rod is a complex comprising several units. Figure 2.11 schematically shows a connecting rod made up of six main parts: the connecting rod body, cap, two shells and two screws. The latter and related mechanical actions are not shown in the diagram. To function normally, the screws need to be sufficiently tightened and the load exerted by the crank pin through the lubricant film needs to be sufficiently low for the system to behave as a solid with an elastic behavior. In this case, the elastic and thermal compliance matrices described in Chapter 4 in [BON 14b] and Chapters 4 and 5 in [BON 14a] used in the linear model are valid.

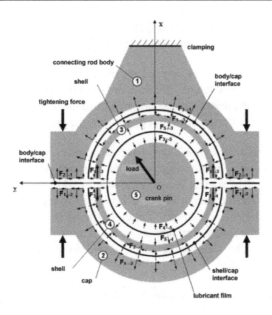

Figure 2.11. *Diagram of the big end bearing for a multibody connecting rod*

However, if the screws are not tight enough and/or the load applied is too high, slippages and openings, in the most severe cases, can occur at the interfaces of the complex-forming solids. A more complex model, based on research related to interface contact interactions, is required to take these possibilities into consideration.

2.4.1. *Interfaces and interactions: main assumptions*

The diagram in Figure 2.11 identifies the interfaces and the corresponding interactions (except those related to screws):

– the interface between the connecting rod body and the cap, located at the joint plane. This plane does not necessarily coincide with the plane O x y (some connecting rods have an oblique plane joint). In the following, non-planar interfaces (obtained by rod breaking or grooved interfaces) are excluded. Main hypothesis for this interface: unilateral contact with friction;

– the interface between the connecting rod body and shell 3, cylindrical: unilateral contact with friction;

– the interface between the connecting rod cap and shell 4, cylindrical: unilateral contact with friction;

– the interface between shell 3 and shell 4, located at the joint plane: unilateral contact resulting in shrinking (the shells are slightly larger than the housing).

Different screw models can be used for the complex[1]. The threaded part of the rod body can be shorter or longer: if it does not extend beyond the joint plane, it may have a fitted part at the joint plane serving as a centering pin between the cap and the rod body. In this case, movement is not possible in the joint plane around the screw.

In contrast, the screw shank can have a smaller diameter than the bore in the body and cap over a length, which extends from one side to the other side of the joint plane. The lack of contact between the screw and its housing at the joint plane makes it possible for movement to occur between the cap and the connecting rod body. The following descriptions are limited to the latter configuration. This is why the interfaces and the mechanical actions between the screws and the rest of the rod are not mentioned. Only the clamping is taken into account in the form of four balanced forces.

Other mechanical actions acting on the solid parts of the rod are as follows:

– clamping forces assumed to be identical for both screws;

– action of the lubricant film on the shells 3 and 4: normally viscous friction (due to contact actions in the case of mixed lubrication) is assumed to be sufficiently low compared to the other actions to be ignored; given the usual assumptions regarding lubrication (variations in pressure across the film thickness are insignificant), the action equals to the one of the crank pin on the film;

– the action of the piston on the connecting rod, represented by action at the cut plane of the rod.

When tightening the screws during assembly, a contact pressure is created on the backs of the shells and at the interface between the shells. It is assumed that these contact actions are known: as the bearing is loaded with hydrodynamic pressure, the actions will be modified and the criteria to be satisfied for controlling the sliding movements or the openings consider mechanical actions in their totality and not just their variations. It is assumed that the bearing shape is the one obtained after the screw tightening and at a reference temperature of 20°C.

---

[1] For some rods, the screws are replaced with bolts whose nut is placed on the side of the cap.

2.4.2. Equations of unilateral contact with friction and equilibrium equations

Figure 2.12 shows the typical configuration of a unilateral contact.

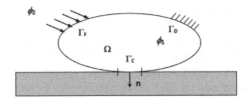

Figure 2.12. *Elastic body in contact with a rigid body*

The unilateral contact problem is based on a strict description of the condition avoiding penetration of a deformable solid that is in contact with a rigid obstacle[2]. It is assumed that the deformable solid occupies the domain Ω of the border Γ sufficiently regularly separated into $\Gamma_F \cup \Gamma_D \cup \Gamma_C$ where Γ_C is the initial contact area for the solid and the obstacle, defined *a priori* with the maximum size of the possible contact area (the interfaces are identified in Figure 2.12).

At Γ_C, the displacement **u** and the contact force **F** are both separated into normal and radial components as:

$$\mathbf{u} = u_N \mathbf{n} + \mathbf{u}_T \qquad [2.2]$$

$$\mathbf{F} = F_N \mathbf{n} + \mathbf{F}_T \qquad [2.3]$$

where $\mathbf{u}_T$ and $\mathbf{F}_T$ are vectors belonging to the tangent plane and u_N and F_N are algebraic measures following the direction of the normal. **n** describes the normal outward of Γ_C.

Unilateral contact conditions are described as:

$$\begin{cases} u_N \leq 0 \\ F_N \leq 0 \\ u_N F_N = 0 \end{cases} \qquad [2.4]$$

These equations well characterize unilateral contact:

– no penetration of the solid in the obstacle ($u_N \leq 0$);

---

[2] If both solids are deformable, the formulation remains the same, the relative displacement takes into account the deformations of both solids.

– normal force exerted by an obstacle on the solid is a compression ($F_N \leq 0$);

– the complementary relationship $u_N F_N \leq 0$ well describes the two possible situations: (1) contact is established ($u_N = 0$), the normal force is active ($F_N < 0$); (2) separation ($u_N < 0$), the force must be zero ($F_N = 0$).

For modeling unilateral contact with friction, Coulomb's law of friction will be used. It is the oldest and best-known law describing friction. It has the two main basic components: the threshold and the only normal stress dependence (no influence of the contact area on the friction coefficient). It can be written as:

$$\|\mathbf{F}_T\| \leq f|F_N| \quad \text{with} \quad \begin{cases} \text{if } \|\mathbf{F}_T\| < f|F_N| \text{ then } \dot{\mathbf{u}}_T = \mathbf{0} \\ \text{if } \|\mathbf{F}_T\| = f|F_N| \text{ then } \exists \lambda > 0 \text{ such that } \dot{\mathbf{u}}_T = -\lambda \mathbf{F}_T \end{cases} \quad [2.5]$$

In addition to the equations described in Chapter 2 of [BON 14b], the system for solving the TEHD problem of a multibody rod big end bearing must integrate these equations for the previously mentioned interfaces: rod body–cap, body–shell 3, cap–shell 4 and shell 3–shell 4. Simplified setups can be considered depending on whether only the joint plane or the shell shrinking is examined.

The system should also integrate the static equilibrium equations for each solid, except for the connecting rod body that is assumed to be in equilibrium due to clamping forces at the cut level. Assuming the tightening forces are identical on the two screws (O moments are balanced), the following equilibrium equations can be written:

cap: $\mathbf{F}_{1\to 2} + \mathbf{F}_{4\to 2} + \mathbf{F}_{tightening \to 2} = \mathbf{0}$; $\mathbf{M}_O(\mathbf{F}_{1\to 2}) + \mathbf{M}_O(\mathbf{F}_{4\to 2}) = \mathbf{0}$

shell 3: $\mathbf{F}_{1\to 3} + \mathbf{F}_{4\to 3} + \mathbf{F}_{film \to 3} = \mathbf{0}$; $\mathbf{M}_O(\mathbf{F}_{1\to 3}) + \mathbf{M}_O(\mathbf{F}_{4\to 3}) = \mathbf{0}$

shell 4: $\mathbf{F}_{2\to 4} + \mathbf{F}_{3\to 4} + \mathbf{F}_{film \to 4} = \mathbf{0}$; $\mathbf{M}_O(\mathbf{F}_{2\to 4}) + \mathbf{M}_O(\mathbf{F}_{3\to 4}) = \mathbf{0}$

The moments at O created by the actions exerted by the hydrodynamic pressure field are assumed to be insignificant.

2.4.3. *Compliance matrices*

To determine elastic deformations of the connecting rod body, cap or shells due to mechanical actions (hydrodynamic pressure field in the film, normal and radial stress fields on the interfaces between the solids, screw tightening forces), a set of compliance matrices are precalculated using the technique described in Chapter 4 of [BON 14b].

For each set of solids $S_i \cup S_j$, the concerned areas are identified. For example, for the connecting rod body–shell 3 set ($S_1 \cup S_3$, Figure 2.13):

– the cylindrical surface of the shell that endures the pressure forces exerted by the lubricant film and the contact forces with the shaft in the case of mixed lubrication, denoted by $\mathbf{F}_{f \to 3}$;

– the surface of the joint plane that endures the actions $\mathbf{F}_{2 \to 1}$ exerted by the cap (solid S_2) on the connecting rod body and $\mathbf{F}_{4 \to 3}$ exerted by the shell 4 on shell 3;

– the bearing surfaces of the screw heads that support the clamping force.

The connecting rod body is held in position by clamping at the cut plane. After combining the finite elements and taking into consideration boundary conditions (clamping at cut level, plane of symmetry at mid-thickness in the case of a symmetrically loaded symmetric rod), the compliance matrices can be obtained by applying a normal unitary force (and possibly in the second phase a tangential unitary force) on each of the surface nodes and recording the normal and possibly the tangential displacement of each of the nodes in the same surface but also of each of the nodes on the other surfaces.

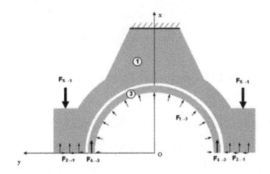

Figure 2.13. *Mechanical actions applied to the "con rod body–shell 3" system*

In the connecting rod body–shell 3 system ($S_1 \cup S_3$ denoted as S_{13}), by identifying two surfaces, it is possible to obtain four compliance matrices, which can be denoted as $C_{24 \to 13}^{13-24}, C_{24 \to 13}^{3-f}, C_{f \to 13}^{13-24}$ and $C_{f \to 13}^{3-f}$, plus two vectors corresponding to normal and radial displacement of the two interfaces produced by clamping forces $\mathbf{F}_S$: $C_{FS \to 13}^{13-24}$ and $C_{FS \to 13}^{3-f}$.

In the notation $C_{k \to l}^{i-j}$, i–j refers to the interface where the displacement is recorded, the interface between the solid or group of solids i and the film f or the group of solids j. k refers to the solid or group of solids that exerts the action and l to the solid or group of solids that receives it. Each matrix C can be composed of four submatrices $C_{k \to l\ N}^{i-j\ N}$, $C_{k \to l\ T}^{i-j\ N}$, $C_{k \to l\ N}^{i-j\ T}$ and $C_{k \to l\ T}^{i-j\ T}$ where N and T refer, respectively, to the normal and tangential displacement or force. For the tangential forces, due to the bearing symmetry relative to the plane O **xy**, it is considered that the axial component plays an insignificant role and can therefore be ignored.

Regarding the tangential component, the same applies for the displacement created by different elemental forces applied.

For the actions performed by the film and the tightening forces, the tangential forces are either insignificant or null. For displacements at the film–shell interface, only the normal component is recorded.

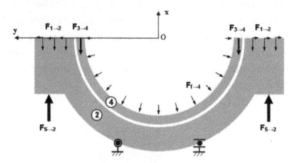

Figure 2.14. *Connecting rod cap in static equilibrium*

For the cap (Figure 2.14), there are also four compliance matrices $C_{13 \to 24}^{13-24}$, $C_{13 \to 24}^{4-f}$, $C_{f \to 24}^{13-24}$ and $C_{f \to 24}^{4-f}$ and two vectors corresponding to the actions of the tightening forces $C_{FS \to 24}^{13-24}$ and $C_{FS \to 24}^{4-f}$. Unlike the connecting rod body, the cap, once isolated, is not maintained in position by any clamping. To apply the unitary forces at each node interface, a finite element model must be applied while maintaining isostatic constraints. Because of reactions at the node level concerned with these constraints, each elementary solution will show deformations in the area surrounding these nodes. However, this will not affect the final deformation, as the static equilibrium equations are integrated in the complete system of equations related to the TEHD problem. Figure 2.14 shows an example of the isostatic constraints and the group of actions that affect the cap and give a resultant torsor of zero. Because of the symmetry of the model, only the part located in the space $z \geq 0$ will be examined with an imposed displacement of zero in the plane $z = 0$.

For the set shell 3–shell 4 (Figure 2.15), there are nine compliance matrices $C_{12 \to 34}^{12-34}$, $C_{f \to 34}^{12-34}$, $C_{12 \to 34}^{34-f}$ and $C_{f \to 34}^{34-f}$. At the lubricant film boundary, action is assumed to only be normal and the displacement is only recorded by its normal component. Similarly as for the rod cap–shell 4 complex, isostatic constraints must be applied to the two-shell set, and similarly as before this will not affect the final deformation, as the static equilibrium equations are integrated in the complete system of equations related to the TEHD problem. Figure 2.15 shows the shell 3–shell 4 set with isostatic constraints and all the actions that it endures and that give a resultant torsor of zero.

It must be noted that the static equilibrium equation relative to the moments around the O z axis requires that moment of the tangential force on the back side of the shells is zero with respect to the same axis. The diagram shown in Figure 2.15 does not mention button stop for locking the shell rotation[3].

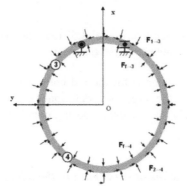

Figure 2.15. *Shell system in static equilibrium*

Figure 2.16. *Shell mesh (half part $z \geq 0$)*

---

3 Certain shells do not have a button stop or dowel pin. In this case the shell is maintained only by its shrinking in the housing.

Figure 2.17. *Radial deformation of connecting rod body housing: Unitary radial force applied to housing mid-plane center. Unitary normal force applied to left part of joint plane*

Because of the low thickness of the shells, it can be assumed that the node displacements on both sides of the shells are the same no matter the side on which the unitary force is applied (elastic shell assumptions). In this case, it is desirable that the mesh on both sides of the shells be identical (Figure 2.16). In this case, there is only a single compliance matrix that allows displacement of the back side of the shells.

The crank pin compliance matrix (solid 5) remains the same as defined for the single-body model of the connecting rod (see section 2.3).

Elementary solutions obtained for the rod body–shell 4 and for the cap–shell 4 complexes, in addition to the large local deformation related to the area where the unitary force is applied, show a notable housing opening due to its semicylindrical shape (Figure 2.17).

2.4.4. *Finite element modeling of the contact in the joint plane*

The problem of unilateral contact with friction can be broken down into a normal and a tangential problem. The normal contact problem consists of determining the open and closed areas and calculating the normal stress field on the contact area and the normal displacements. The tangential problem determines the adhesion and sliding areas and calculates the tangential stresses and relative sliding. Thus, in the case of finite element modeling, for the normal problem, we research at each step the nodes that are capable of penetrating the opposing side. For each of these nodes, a penalty term is introduced into the model to move the node on the surface of the penetrated body back. This penalty term is based on the penetration distance. This term can be seen as a spring with a high degree of stiffness connecting the examined node with its projection on the surface of the antagonist solid (Figure 2.18).

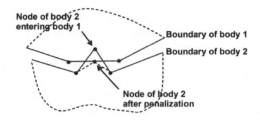

Figure 2.18. *Penalty technique for the contact management*

The body and the connecting rod cap are meshed independently. The combination of both parts results in a situation at the joint plane where the nodes on both sides are not necessarily face to face. If this is the case, it requires the examination of the projections of one surface on the other to take into account these interactions.

Figure 2.19 shows a diagram of the contact problem at the joint plane. The two solids in contact consist of the rod body and the shell 3, on the one hand, and the cap and shell 4, on the other hand. The mechanical actions taken into consideration are shown in the figure. The interactions between the rod body and shell 3 and between the cap and shell 4 are internal actions that are not involved in this step. *In the remainder of section 2.4.4, "body" will refer to the rod body and its shell and "cap" will refer to the cap and its shell. Sets 1–3 will thus simply be denoted as 1 and sets 2–4 will be denoted as 2.*

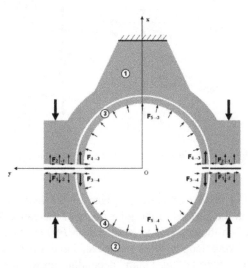

Figure 2.19. *Contact problem for the joint plane*

The surfaces corresponding to the joint plane have two-dimensional (2D) meshes from the finite element three-dimensional (3D) mesh of the body, on the one hand, and of the cap, on the other hand. As the 3D meshes of the two parts of the rod are independent from each other, the 2D meshes are different and the nodes on each surface are not orientated facing one another. When addressing contact problems, with or without friction, one of the two surfaces should be considered as the "master" and the other one as the "slave": the nodes on the slave surface are then projected onto the master surface [ZIE 00]. For the model presented, developed by Tran [TRA 06], both surfaces are successively considered as master surfaces: one double projection is formed so as not to give a leading role to any of the surfaces in order to increase the precision of the contact problem solution. Obviously, the model is simplified a lot when the two surface meshes are identical: the projections are no longer necessary as both surfaces carry out the same role.

2.4.4.1. *Normal problem or opening problem*

The normal problem is treated separately for the body and the cap. The respective numbers of nodes in the meshes of these solids are given by $nnpj_1$ and $nnpj_2$. The contact is defined between the nodes of one of the two surfaces and their projections on the mesh of the other surface.

It is assumed that the tangential forces in the joint plane and that the hydrodynamic pressure forces at film level are known.

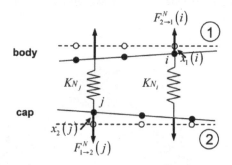

Figure 2.20. *Normal contact in the joint plane: notations*

In Figure 2.20, a node i of the joint plane of the body side (shown as 1) is projected on $P(i)$ on the surface of the joint plane of the cap side (shown as 2) and conversely a node j on the cap side is projected on $P(j)$ on the surface of the body. Penalty springs, of zero length when unloaded, have stiffness K_N in the normal direction that is either zero or "very high" depending on whether the corresponding node is in an "open" or "closed" state. $x_1(i)$ and $x_2(j)$ represent, respectively, the normal movement of the nodes i and j. $F^N_{2\to1}(i)$ and $F^N_{1\to2}(j)$ are, respectively, the normal forces exerted on i and j on the body by the first and on the cap by the second.

2.4.4.1.1. The cap is the "master" body

The node i of surface 1 is projected on $P(i)$ on surface 2. The algebraic measure $F^N_{2\to1}(i)$ of the nominal force of the node i can be written as (Figure 2.20):

$$F^N_{2\to1}(i) = -K_{N_i}\Delta x_i = -K_{N_i}\left[x_1(i) - x_2(P(i))\right] \qquad [2.6]$$

where $x_2(P(i))$ is the normal displacement of point $P(i)$.

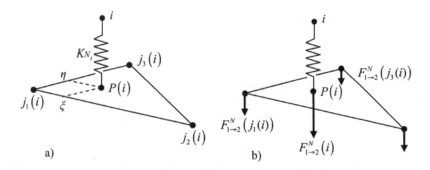

Figure 2.21. *Normal contact in the joint plane: modelization*

The normal movement of node i on surface 1 is given by:

$$x_1(i) = \sum_{k=1}^{nnpj_1} C^{12}_{2\to1}{}^N_N(i,k) F^N_{2\to1}(k) + \sum_{k=1}^{nnpj_1} C^{12}_{2\to1}{}^N_T(i,k) F^T_{2\to1}(k)$$
$$+ \sum_{k=1}^{nnf_3} C^{12}_{f\to3}{}^N(i,k) F_f(k) + C^{12}_{FS\to1}{}^N(i) F_S \qquad [2.7]$$

where nnf_3 is the number of nodes in the mesh of the surface of shell 3 in contact with the film.

If it is assumed that the surface mesh consists of linear triangles[4], the displacement of the point $P(i)$, projection of i on surface 2, can be obtained by interpolation using the displacements $j_1(i)$, $j_2(i)$ and $j_3(i)$ of the nodes of the surface 2 triangle containing the projection $P(i)$ of i (Figure 2.21(a)):

$$x_2(P(i)) = \sum_{k=1}^{3} N_k(\xi,\eta) x_2(j_k(i)) \qquad [2.8]$$

where ξ and η are parametric coordinates of $P(i)$.

The normal component $F_{1\rightarrow 2}^{N}(i)$ of the force at point $P(i)$, opposing $F_{2\rightarrow 1}^{N}(i)$, can be divided into three forces related to the nodes $j_1(i)$, $j_2(i)$ and $j_3(i)$ of the triangle (Figure 2.21(b)), which can be written as:

$$\begin{cases} f_1(i) = N_1(\xi,\eta) F_{1\rightarrow 2}^{N}(i) \\ f_2(i) = N_2(\xi,\eta) F_{1\rightarrow 2}^{N}(i) \\ f_3(i) = N_3(\xi,\eta) F_{1\rightarrow 2}^{N}(i) \end{cases}$$

where N_k represents the interpolation functions in the surface element such that $N_1 + N_2 + N_3 = 1$. At node l of surface 2, these forces accumulate and are expressed as:

$$F_{1\rightarrow 2}^{N}(l) = \sum_{\substack{m=1 \\ P(m)\in E_l}}^{nnpj_1} f_{k(l)}(m) = \sum_{\substack{m=1 \\ P(m)\in E_l}}^{nnpj_1} N_{k(l)}(\xi_m,\eta_m) F_{1\rightarrow 2}^{N}(m) \qquad [2.9]$$

The contribution comes only from node m whose projection $P(m)$ is located in one of the elements connected to node l; $k(l)$ is then the node l number in the element.

The normal displacement of node j on surface 2 can be written as:

$$x_2(j) = \sum_{l=1}^{nnpj_2} C_{1\rightarrow 2}^{12}{}^{N}(j,l) F_{1\rightarrow 2}^{N}(l) + \sum_{l=1}^{nnpj_2} C_{1\rightarrow 2}^{12}{}^{N}(j,l) F_{1\rightarrow 2}^{T}(l)$$
$$+ \sum_{l=1}^{nnf_4} C_{f\rightarrow 4}^{12}{}^{N}(j,l) F_f(l) + C_{FS\rightarrow 2}^{12}{}^{N}(j) F_S + Ca - y(j)\alpha$$

---

[4] The transformation of the equations in the case of different element types does not pose difficulties.

where nnf_4 represents the number of nodes in the mesh on the surface of shell 4 in contact with the film.

$Ca - y(j)\alpha$ represents a rigid body displacement of the rod cap, in which C is the radial displacement, a is the dimensionless displacement in the x direction of a reference point in the cap, $y(j)$ is the ordinate of node j and α is the rigid body rotation. The introduction of this rigid body displacement is necessary for placing the joint plane of the cap in a position compatible with the rod body joint plane.

By replacing $F_{1\to 2}^N(l)$ using equation [2.9], the following can be obtained:

$$x_2(j) = \sum_{l=1}^{nnpj_2} C_{1\to 2}^{12} {}^N\!(j,l) \sum_{\substack{m=1 \\ P(m)\in E_l}}^{nnpj_1} N_{k(l)}(\xi_m, \eta_m) F_{1\to 2}^N(m) + \sum_{l=1}^{nnpj_2} C_{1\to 2}^{12} {}^N\!(j,l) F_{1\to 2}^T(l)$$

$$+ \sum_{l=1}^{nnf_4} C_{f\to 4}^{12} {}^N\!(j,l) F_f(l) + C_{FS\to 2}^{12} {}^N\!(j) F_S + Ca - y(j)\alpha$$

and finally:

$$x_2(P(i)) = \sum_{k=1}^{3} N_k(\xi,\eta) \Bigg[\sum_{l=1}^{nnpj_2} C_{1\to 2}^{12} {}^N\!(j_k(i),l) \sum_{\substack{m=1 \\ P(m)\in E_l}}^{nnpj_1} N_{k(l)}(\xi_m,\eta_m) F_{1\to 2}^N(m)$$

$$+ \sum_{l=1}^{nnpj_2} C_{1\to 2}^{12} {}^T\!(j_k(i),l) F_{1\to 2}^T(l) + \sum_{l=1}^{nnf_4} C_{f\to 4}^{12} {}^N\!(j_k(i),l) F_f(l) \qquad [2.10]$$

$$+ C_{FS\to 2}^{12} {}^N\!(j_k(i)) F_S + Ca - y(j_k(i))\alpha \Bigg]$$

As it is known that:

$$y(i) = y(P(i)) = \sum_{k=1}^{3} N_k(\xi,\eta) y(j_k(i)) \;;\; \sum_{k=1}^{3} N_k(\xi,\eta) = 1$$

equation [2.10] can also be rewritten as:

$$x_2(P(i)) = \sum_{k=1}^{3} N_k(\xi,\eta)\Bigg[\sum_{l=1}^{nnpj_2} C_{1\to 2}^{12}{}_N^N(j_k(i),l) \sum_{\substack{m=1 \\ P(m)\in E_l}}^{nnpj_1} N_{k(l)}(\xi_m,\eta_m) F_{1\to 2}^N(m)$$

$$+ \sum_{l=1}^{nnpj_2} C_{1\to 2}^{12}{}_T^N(j_k(i),l) F_{1\to 2}^T(l) + \sum_{l=1}^{nnf_4} C_{f\to 4}^{12}{}^N(j_k(i),l) F_f(l) \qquad [2.11]$$

$$+ C_{FS\to 2}^{12}{}^N(j_k(i)) F_S \Bigg] + Ca - y(i)\alpha$$

By putting the equations for $x_1(i)$ and $x_2(P(i))$ given by equations [2.7] and [2.11] in equation [2.6], it is possible to construct a system of equations for the situation when the cap is considered as the master solid.

2.4.4.1.2. The connecting rod body is the "master" solid

The node j of surface 2 is projected on $P(j)$ of surface 1. The algebraic measure $F_{1\to 2}^N(j)$ of the normal force on node j of the cap (Figure 2.14) can be written as:

$$F_{1\to 2}^N(j) = -K_{N_j}\Delta x_j = -K_{N_j}\big[x_2(j) - x_1(P(j))\big] \qquad [2.12]$$

where $x_2(j)$ is the normal displacement of node j on the mesh of surface 2, $x_1(P(j))$ is the normal displacement of $P(j)$ and K_{N_j} is the penalty coefficient similar to a normal stiffness at node j.

Similar rearrangements like those carried out for the connecting rod body result in the following expressions for $x_2(j)$ and $x_1(P(j))$:

$$x_2(j) = \sum_{k=1}^{nnpj_2} C_{1\to 2}^{12}{}_N^N(j,k) F_{1\to 2}^N(k) + \sum_{k=1}^{nnpj_2} C_{1\to 2}^{12}{}_T^N(j,k) F_{1\to 2}^T(k)$$

$$+ \sum_{k=1}^{nnf_4} C_{f\to 4}^{12}{}^N(j,k) F_f(k) + C_{FS\to 2}^{12}{}^N(j) F_S + Ca - y(j)\alpha$$

$$x_1(P(j)) = \sum_{k=1}^{3} N_k(\xi,\eta)\Bigg[\sum_{l=1}^{nnpj_1} C_{2\to 1}^{12}{}_N^N(i_k(j),l) \sum_{\substack{m=1 \\ P(m)\in E_l}}^{nnpj_2} N_{k(l)}(\xi_m,\eta_m) F_{2\to 1}^N(m)$$

$$+ \sum_{l=1}^{nnpj_1} C_{2\to 1}^{12}{}_T^N(i_k(j),l) F_{2\to 1}^T(l) + \sum_{l=1}^{nnf_3} C_{f\to 3}^{12}{}^N(i_k(j),l) F_f(l)$$

$$+ C_{FS\to 1}^{12}{}^N(i_k(j)) F_S \Bigg]$$

Substituting these expressions into [2.12], a system of equations can be obtained related to the normal problem when the body is considered as the master solid.

2.4.4.1.3. Closing the equation system for the normal contact problem

To the system of equations [2.6] and [2.12], we must associate the cap equilibrium equations that include screw tightening forces and the set of components in the x direction of the resultant of the hydrodynamic pressure forces acting on shell 4 and the nodal forces at the joint plane:

$$F_S + \sum_{i=1}^{nnpj_1} F_{1\to 2}^N(i) + \sum_{j=1}^{nnpj_2} F_{1\to 2}^N(j) + W_{f\to 4}^x = 0 \qquad [2.13]$$

with:

$$W_{f\to 4}^x = R\iint_{\Omega_4} p\cos\theta\, d\theta dz = R\int_{-L/2}^{L/2}\int_{-\pi/2}^{\pi/2} p\cos\theta\, d\theta dz$$

When the forces of viscous friction are assumed to be insignificant, the balance of moments about the O z axis of the bearing (Figure 2.13) of the forces applied to the cap can be written as:

$$\sum_{i=1}^{nnpj_1} F_{1\to 2}^N(i)y(i) + \sum_{j=1}^{nnpj_2} F_{1\to 2}^N(j)y(j) = 0 \qquad [2.14]$$

When the forces of viscous friction are taken into account, the above equation becomes:

$$\sum_{i=1}^{nnpj_1} F_{1\to 2}^N(i)y(i) + \sum_{j=1}^{nnpj_2} F_{1\to 2}^N(j)y(j) + C_{f\to 4} = 0$$

where $C_{f\to 4}$ represents the friction torque created by the film fluid on the surface where the shell connects to the cap (solid 4):

$$C_{f\to 4} = R\iint_{\Omega_4} \tau\, d\theta dz = R\int_{-L/2}^{L/2}\int_{-\pi/2}^{\pi/2} \tau\, d\theta dz$$

Equations [2.6] and [2.12]–[2.14] constitute a system of $nnpj_1 + nnpj_2 + 2$ equations corresponding to $nnpj_1 + nnpj_2 + 2$ unknowns that are the nodal forces $F_{1\to 2}^N(i)$ applied to $nnpj_1$ nodes at the junction plane for the body side, the nodal forces $F_{1\to 2}^{N}(j)$ applied to $nnpj_2$ nodes at the junction plane for the cap side and the two parameters a and α describing the cap position.

2.4.4.1.4. Algorithm for solving the normal problem

The solution of the normal problem consists of determining whether a point at the interface is open or closed: an open point must satisfy the condition of non-interpenetration and a closed point must satisfy the condition of normal compression force of one solid on the other.

For a given node state, the system of linear equations [2.6] and [2.12]–[2.14] can be resolved. This gives new normal force fields $F_{1\to 2}^{N}(i)$ and $F_{1\to 2}^{N}(j)$ and their openings can be described by the following equations:

$$\begin{cases} \text{opening}(i) = -\dfrac{F_{1\to 2}^{N}(i)}{K_N(i)} \quad ; \quad i = 1, nnpj_1 \\ \text{opening}(j) = -\dfrac{F_{1\to 2}^{N}(j)}{K_N(j)} \quad ; \quad j = 1, nnpj_2 \end{cases} \quad [2.15]$$

The status of all points not satisfying the conditions relative to their state will be changed. This results in a change in the contact area. This iterative process of the normal problem is engaged and the convergence is achieved when no status changes are registered.

The normal problem solution gives:

– the distribution of opening and closing zones (contact areas);

– the distribution of normal nodal forces;

– the node openings at the joint plane;

– the coefficient a and the angle α of rigid body displacement of the cap.

2.4.4.2. *Tangential problem or stick–slip problem*

The approach is the same as the one used for the normal problem. The tangential problem is approached separately for the body and for the cap. It is assumed that the contact area and the maximum force on the joint plane as well as the hydrodynamic pressure forces at the film level are known.

The displacement of the nodes in axial direction z is assumed to be insignificant.

When the cap is the "master" solid, the radial force of node i of solid 1 can be written as:

$$\left|F_{2\to1}^{T}(i)\right| - f\left|F_{2\to1}^{N}(i)\right| = K_{T_i}\left[y_1(i) - y_2(P(i)) - \overline{y_1(i)} + \overline{y_2(P(i))}\right] \quad [2.16]$$

where $F_{1\to2}^{N}(i)$ and $F_{1\to2}^{T}(i)$ are the normal and tangential components of the force on node i, $y_1(i)$ and $\overline{y_1(i)}$ are the tangential displacements of node i, respectively, for the current and the previous load, $y_1(P(i))$ and $\overline{y_1(P(i))}$ are the tangential displacements of the node i projection on surface 2, respectively, for the current and the previous load and K_{T_i} is the penalty coefficient similar to a tangential stiffness at node i.

K_{T_i} is "big" if the node sticks and "small" if the node slides. In the second case, the following equation [2.16] can be written as:

$$\left|F_{2\to1}^{T}(i)\right| - f\left|F_{2\to1}^{N}(i)\right| = 0$$

This corresponds to the sliding condition given by equation [2.5].

The friction coefficient f equals the adhesion coefficient f_S (or static friction) when the point is adherent or equals the dynamic friction coefficient f_D if the point slides. The last is always less than the adhesion coefficient.

It is worth noting that unlike the normal problem, the solution to the tangential problem depends on the load history.

Following the same approach as with the normal components, the following expressions can be written as:

$$y_1(i) = \sum_{k=1}^{nnpj_1} C_{2\to1\,N}^{12\ T}(i,k) F_{2\to1}^{N}(k) + \sum_{k=1}^{nnpj_1} C_{2\to1\,T}^{12\ T}(i,k) F_{2\to1}^{T}(k)$$
$$+ \sum_{k=1}^{nnf_3} C_{f\to3}^{12\ T}(i,k) F_f(k) + C_{FS\to1}^{12\ T}(i) F_S$$

$$y_2(P(i)) = \sum_{k=1}^{3} N_k(\xi,\eta) \Big[\sum_{l=1}^{nnpj_2} C_{1\to2}^{12}{}_N^T(j_k(i),l) \sum_{\substack{m=1 \\ P(m)\in E_I}}^{nnpj_1} N_{k(l)}(\xi_m,\eta_m) F_{1\to2}^T(m)$$

$$+ \sum_{l=1}^{nnpj_2} C_{1\to2}^{12}{}_T^T(j_k(i),l) F_{1\to2}^T(l) + \sum_{l=1}^{nnf_4} C_{f\to4}^{12}{}^T(j_k(i),l) F_f(l)$$

$$+ C_{FS\to2}^{12}{}^T(j_k(i)) F_S \Big] + Cb$$

where Cb represents the y displacement component of the rigid solid at the cap. It must be noted that the displacement along y does not depend on the rotation of the rigid solid at the cap as the joint plane is located at $x = 0$.

By inserting these expressions into equation [2.15], the first system of equations related to the tangential problem can be created. The unknowns are the tangential components of contact forces at the current step and the coefficient b.

When the connecting rod body is the "master solid", the radial force on node j at solid 2 can be written as:

$$\left| F_{1\to2}^T(j) \right| - f \left| F_{1\to2}^N(j) \right| = K_{T_j} \Big[y_2(j) - y_1(P(j)) - \overline{y_2(j)} + \overline{y_1(P(j))} \Big] \quad [2.17]$$

with:

$$y_2(j) = \sum_{k=1}^{nnpj_2} C_{1\to2}^{12}{}_N^T(j,k) F_{1\to2}^N(k) + \sum_{k=1}^{nnpj_2} C_{1\to2}^{12}{}_T^T(j,k) F_{1\to2}^T(k)$$

$$+ \sum_{k=1}^{nnf_4} C_{f\to4}^{12}{}^T(j,k) F_f(k) + C_{FS\to2}^{12}{}^T(j) F_S + Cb$$

$$y_1(P(j)) = \sum_{k=1}^{3} N_k(\xi,\eta) \Big[\sum_{l=1}^{nnpj_1} C_{2\to1}^{12}{}_N^T(i_k(j),l) \sum_{\substack{m=1 \\ P(m)\in E_I}}^{nnpj_2} N_{k(l)}(\xi_m,\eta_m) F_{2\to1}^T(m)$$

$$+ \sum_{l=1}^{nnpj_1} C_{2\to1}^{12}{}_T^T(i_k(j),l) F_{2\to1}^T(l) + \sum_{l=1}^{nnf_3} C_{f\to3}^{12}{}^T(i_k(j),l) F_f(l)$$

$$+ C_{FS\to1}^{12}{}^T(i_k(j)) F_S \Big]$$

To the system of equations [2.16] and [2.17] written for all nodes at the junction plane with the closed status is associated the cap equilibrium equation in the y direction with the combination of the y components of forces acting on it.

$$\sum_{i=1}^{nnpj_1} F_{1\to 2}^T(i) + \sum_{j=1}^{nnpj_2} F_{1\to 2}^T(j) + W_{f\to 4}^y = 0 \qquad [2.18]$$

with:

$$W_{f\to 4}^y = R\iint_{\Omega_4} p\sin\theta \, d\theta dz = R\int_{-L/2}^{L/2}\int_{-\pi/2}^{\pi/2} p\sin\theta \, d\theta dz$$

2.4.4.2.1. Algorithm for solving the tangential problem

It must be determined whether a closed point is adherent or sliding. This can be done by analyzing the two following criteria:

– a closed adherent point must satisfy the inequality of the threshold or adhesion limit;

– a closed sliding point must produce negative work and thus the tangential displacement and the tangential force must be in opposite directions.

After solving the normal problem and obtaining the status of the closed nodes, the linear system consisting of equations [2.16]–[2.18] can be solved. This gives new tangential forces $F_{1\to 2}^T(i)$ and $F_{1\to 2}^T(j)$, and for nodes with the "slip" status, the relative slips are given by the expressions:

$$\begin{cases} sliping(i) = -\dfrac{\left(\left|F_{1\to 2}^T(i)\right| - f\left|F_{1\to 2}^N(i)\right|\right)}{K_T(i)} \quad ; \quad i=1, nnpj_1 \text{ and status}(i) = slip \\ sliping(j) = -\dfrac{\left(\left|F_{1\to 2}^T(j)\right| - f\left|F_{1\to 2}^N(j)\right|\right)}{K_T(j)} \quad ; \quad j=1, nnpj_2 \text{ and status}(j) = slip \end{cases}$$

The above relations applied to nodes with the status "stick" ($K_T(i)$ very large) allow us to identify the direction of the slip "risk", thus lifting the ambiguity about the sign of the tangential force, and to get rid of the absolute values in equations [2.16] and [2.17].

In the same way as for the normal problem, any point that does not satisfy the appropriate conditions changes its status. The tangential problem convergence is required for the stabilization of adherence and sliding areas.

Solving the tangential problem gives:

– the distribution of adhesion and sliding zones;

– the distribution of nodal radial forces;

– the node sliding;

– the coefficient b for the rigid body displacement of the cap.

2.4.4.3. *Resolution algorithm for the joint plane behavior problem*

Each of the problems defined in the previous sections uses the results of the other problem as:

– the determination of the closed area and the calculations of normal forces use the tangential forces;

– the determination of the adherence area within the closed area and the calculations of tangential forces use the normal forces.

These two problems are solved alternately until the stabilization of the node statuses occurs.

When the forces in the joint plane are completely known, it is possible to calculate the deformation of the shell surface that is used when calculating the film thickness. For a node i on the surface of shell 3 at the side of the connecting rod body, it can be written as:

$$d_3(i) = \sum_{k=1}^{nnpj_1} C_{2 \to 1\ N}^{3-f}(i,k) F_{2 \to 1}^{N}(k) + \sum_{k=1}^{nnpj_1} C_{2 \to 1\ T}^{3-f}(i,k) F_{2 \to 1}^{T}(k) + \sum_{k=1}^{nnf_3} C_{f \to 3}^{3-f}(i,k) F_f(k) + C_{FS \to 1}^{3-f}(i) F_S \qquad [2.19]$$

For a node j at the surface of shell 4 on the cap, it can be written as:

$$d_4(j) = \sum_{k=1}^{nnph} C_{1\to 2\;N}^{4-f}(j,k) F_{1\to 2}^{N}(k) + \sum_{k=1}^{nnph} C_{1\to 2\;T}^{4-f}(j,k) F_{1\to 2}^{T}(k)$$
$$+ \sum_{k=1}^{nnf_4} C_{f\to 4}^{4-f}(j,k) F_f(k) + C_{FS\to 2}^{4-f}(j) F_S \qquad [2.20]$$
$$+ \left[Ca - y(j)\alpha \right] \cos\theta(j) + \left[Cb + x(j)\alpha \right] \sin\theta(j)$$

The contact algorithm for the joint plane is detailed in Figure 2.22. It must be associated with the general TEHD (or EHD) algorithms defined in Chapter 5 of Volume 3 [BON 14a] as shown in Figure 2.23.

Contact algorithm in the joint plane
With the efforts computed at the precedent call
While the status are not stabilized
    While the open/ close status are not stabilized
        Compute the normal efforts
        For all the nodes with a close status
            If the normal force exerts a traction
                The node is set to the open status
            End if
        else
            if the displacement gives an interpenetration
                The node is set to the close status
            End if
        End for
    End while
    For all the nodes with a close status
        While the stick / slip status are not stabilized
            Compute the tangential efforts
            For all the nodes with a stick status
                If the forces verify |T| > f |N|
                    The node is set to the slip status
                End if
            else
                If the displacement increase products a positive work
                    The node is set to the stick status
                End if
            End for
        End while
    End for
End while
Compute the radial deformation of the sleeve surface

Figure 2.22. *Contact algorithm in the joint plane*

General algorithm with accounting of joint plane contact behavior
Read the data for geometry, mass, rheology, compliance matrices, ...
Read of running parameters and computing hypothesis
Compute the integration matrices [A] and the projection matrices [P] and [P*]
Compute the compliance matrix for the chosen film mesh
Initialization of status in the joint plane at «close» et «slip» for all the nodes
While Values(end of cycle) – Values(beginning of cycle)> ε
    For all the time steps of the cycle
        While the fields for, thickness and temperature are not stabilized
            Resolution of the contact problem for the joint plane (fig. 2.22)
            Resolution of the EHD problem with a Newton-Raphson process
            Find the domain partition into active and non active zones
            If non isothermal assumptions
                Compute the temperature fields
            End if
        End while
        Compute flow rate, friction torque, dissipation energy, ...
        Write the results for the current step
    End for
    The current tangential displacements become the precedent ones
    If the maximum number of cycles is reached
        Stop
    End if
End while

Figure 2.23. *General algorithm with an accounting of the contact in the joint plane*

2.4.4.4. *Example of computation with a 2D model*

Stefani [STE 03] developed a 2D model that considers the openings and the potential shifts in the joint plane. This model was used in calculations describing the big end rod bearing behavior for a high-speed engine. The main data used are presented in Table 2.3.

Crank shaft radius	39.5	mm
Connecting rod length	137	mm
Bearing width	20.85	mm
Crank pin radius	23.5	mm
Radial clearance	0.04	mm
Screw tightening force	44.8	kN
Young's modulus for the con rod	110	GPa
Poisson coefficient for the con rod	0.329	
Lubricant viscosity at ambient pressure	3.53	mPa·s
Piezoviscosity coefficient	3.67	GPa^{-1}
Friction coefficient for the joint plane	0.25	

Table 2.3. *Data for the engine, the bearing and the lubricant*

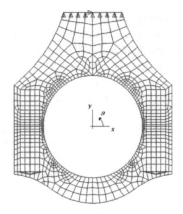

Figure 2.24. *2D mesh for the connecting rod*

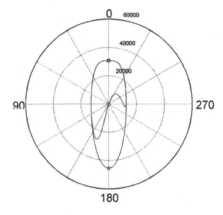

Figure 2.25. *Load diagram (N) at 10,500 rpm*

The 2D profile of the rod and the mesh used is shown in Figure 2.24. The set screws are modeled by adding prestressed beam elements connecting the two extremities of the screws.

The load diagram at 10,500 rpm (Figure 2.25) has a predominantly inertial profile. At an engine angle of 360° (the top dead center at the beginning of the intake phase), the load is at its maximum (approximately 45,000 N) and is orientated facing the rod cap (bearing angle of 180°), with the latter being driven by the piston. At greater turn (angle 0° or 720°), the pressure of the combustion gases completely compensates the inertial effect that gives a load of virtually zero. At angles 180° and 540° (bottom dead center), the push is directed towards the rod.

Figure 2.26 shows the changes in minimal film thickness for both the single-body rod model and the two-body model consisting of the cap and the rod body. In the period between the angles 380° and 430°, a significant difference can be seen between the results of both models. The traction on the rod cap creates an opening at the joint plane and a break in of the bearing continuity profile can be shown in Figure 2.27 that represents the rod deformation at an engine angle of 400°.

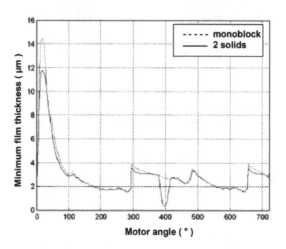

Figure 2.26. *Evolution of the minimum film thickness at 10,500 rpm*

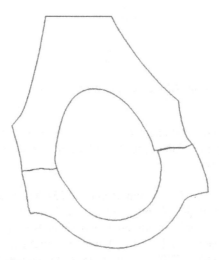

Figure 2.27. *Connecting rod profile at an engine angle of 400°. Deformation amplified by a factor 150*

The minimum thickness calculated at this time is about 0.3 μm and is located at the projecting angle in the right side of the joint plane. The use of chamfers on the shell gives a depth of about 10 μm (lead) allowing the minimum thickness to be increased to an acceptable level [STE 03].

2.4.4.5. *Example of computation with a 3D model*

Tran applied the 3D model presented previously for comparative analysis of the behavior of a rod model made up of photoelastic material placed on a bench reproducing the crank shaft–connecting rod system of a single-cylinder engine. Some of the data and results by Tran [TRA 06] are used for illustrative purposes.

Crank shaft radius	80	mm
Connecting rod length	257.5	mm
Bearing width	20.85	mm
Crank pin radius	20	mm
Radial clearance	0.3	mm
Screw tightening force	125 - 300	N
Young modulus for the connecting rod	3.15	GPa
Poisson coefficient for the connecting rod	0.36	
Lubricant viscosity at 20°C	0.55	Pa s
Thermoviscosity coefficient	0.018	°C^{-1}
Static friction coefficient	0.3	
Dynamic friction coefficient	0.2	

Table 2.4. *Data for the connecting rod, the big end bearing and the lubricant*

The symmetry of the rod and the load allows for the use of a model limited to only half of the space $z \geq 0$. The 3D mesh used is shown in Figure 2.28. The rod body, cap and the tightening screws are all made from the same material (araldite).

Because of experimental contingencies, the rotational frequency is low, which is between 100 and 300 rpm. The presented results correspond to a frequency of 150 rpm.

Given the high flexibility of the material, the load applied by the crank pin on the rod does not exceed 250 N. Figure 2.29 shows the polar load diagrams obtained experimentally and used for calculations. Because of the low rotational frequency, the inertial forces are weak and the traction on the cap at the bottom dead center does not exceed 20 N. To identify all the slippings and openings in the joint plane, the tightening screws were moved outward contrary to their use (Figure 2.28).

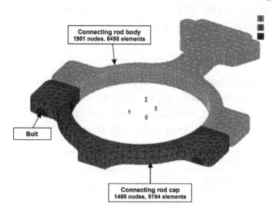

Figure 2.28. *Connecting rod 3D mesh. Half-model due to symmetry*

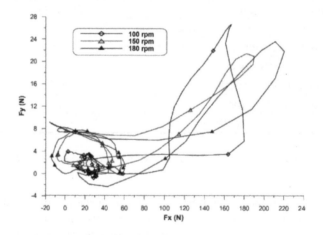

Figure 2.29. *Polar load diagrams*

Figure 2.30 shows the opening and the slipping in the joint plane by a tightening force of 125 N, obtained at a crank shaft angle of 180°. An opening occurs only on the left side (amplitude 1.18 µm). The small number of elements on the joint plane mesh is worth noting. The opening area cannot be precisely defined as it only covers the two strips of elements at the edge of the bearing. For more precise calculations, a finer mesh would be required, with thousands of elements on each side of the joint plane. The complexity of the TEHD algorithm and the high number of needed iterations result in a significant computation time.

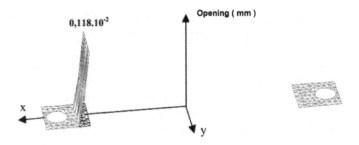

Figure 2.30. *Opening of the joint plane. Crank shaft angle: 180°; screw tightening: 125 N; rotational frequency: 150 rpm*

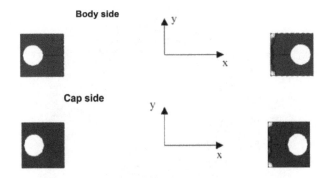

Figure 2.31. *Joint plane slipping. Crank shaft angle: 180°; screw tightening: 300 N; rotational frequency: 150 rpm*

For a screw tightening force of 300 N and the same rotational frequency, the joint plane opening disappears. However, there remains a slippage at the bearing edge as shown in Figure 2.31 with a maximum of 0.5 μm in the corners. The slippage at the side of the body and at the side of the cap slightly differs due to the meshes on the two surfaces not being the same. The double-projection technique used by the model allows this dual representation.

2.4.5. *Modelization of the contact between the housing and the shells*

In this model, it is assumed that the rod body and cap are a single solid denoted by 12. It is also assumed that the shells create a single solid in the form of a cylindrical ring denoted by 34. As the outer cylinder of the latter is larger than the bore shown in set 12 (denoted as "housing"), a prestress shrinking occurs when tightening the screw. Although the shell–bearing contact is considered to have friction, it is assumed that once the rod is assembled and the screws are tightened, the shear stress at the housing–shell interface becomes zero.

The pressure field related to shrinking can be easily obtained by a calculation of finite elements. Elastothermic calculations of the assembled set must be carried out for solids 12 and 34 with the different thermal expansion coefficients α_{12} and α_{34} and by choosing an increase in uniform temperature ΔT such that $(\alpha_{34} - \alpha_{12})D\,\Delta T$ equals the difference between the outer shell diameter and the housing diameter D.

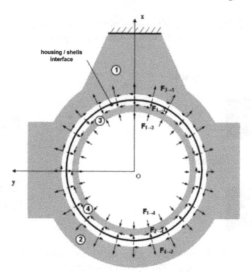

Figure 2.32. *Contact problem between the housing and the shells*

Figure 2.32 shows the mechanical actions taken into account. The shear forces of viscous friction in the film have low amplitude and are considered insignificant. However, their inclusion does not pose any particular difficulties.

The mesh of the solid 1 ∪ 2 is created with standard finite element software, mostly in linear or quadratic tetrahedrons. The particular simple shell shape allows the use of a finite element mesh obtained by simple inward extrusion of the housing surface mesh, as shown in Figure 2.16 where the elements are prisms. For the modeling of the housing–shell contact, the *nnf* nodes at the interface are split and connected by a contact element of two nodes (Figure 2.33). This element has a very high constant normal stiffness if the node has a "closed" status or very low one if the node has an "open" status and a tangential stiffness of zero if the node has an "open" status; very low if the status of the node is "slip" or very high when the status is "stick".

Figure 2.33. *Contact element with friction*

The model is similar to that developed for the join plane with a simplification, due to the coincidence of the nodes of the two bodies in contact. To model the "bimetal" effect that can occur when the shell and housing materials differ (for example the rod is titanium and the shells are mainly steel and copper), the movement of the nodes in the axial direction is not insignificant.

Afterward, the algebraic steps involving radial displacement and radial forces can be obtained by using displacement vector and force projections on the normal orientated from the bearing center toward the outer part of the bearing.

The model developed in this section is designed to be inserted in the general TEHD algorithm (Chapter 5 of [BON 14a]). It is assumed that during its use, the temperature field in the considered solid set is known. The increase in temperature from that assembly is approximately a 100°C and can reach 200°C for the big end rod bearings for F1 engines. It results in significant changes in shell hoop stress, particularly if the materials used are different.

From finite element elastothermic calculations, the normal and tangential stress fields between the housing and the shells are calculated and translated by integrating them into normal, circumreferential and axial nodal forces, respectively, denoted as $F_{ET_{34 \to 12}}^{N}(i)$, $F_{ET_{34 \to 12}}^{C}(i)$ and $F_{ET_{34 \to 12}}^{A}(i)$.

The same calculation gives displacement fields in the axial direction for nodes at the interface of the housing side and the shell side, denoted as $d_{ET_{12}}^{A}(i)$ and $d_{ET_{34}}^{A}(i)$.

2.4.5.1. *Normal problem*

The algebraic measure $F_{34 \to 12}^{N}(i)$ of the normal (or radial) force on node *i* at the housing–shell interface can be given by:

$$F_{34 \to 12}^{N}(i) + F_{ET_{34 \to 12}}^{N}(i) = -K_{N_i} \Delta x_i = -K_{N_i} \left[d_{34}^{N}(i) - d_{12}^{N}(i) \right] \qquad [2.21]$$

where $d_{12}^N(i)$ and $d_{34}^N(i)$ are, respectively, the normal displacements of point i considered as belonging to either the housing or the shells.

The normal displacements $d_{12}^N(i)$ and $d_{34}^N(i)$ of node i can be written as:

$$d_{12}^N(i) = \sum_{k=1}^{nnf} C_{34\to12\ N}^{12-34\ N}(i,k)\left[F_{34\to12}^N(k)+FET_{34\to12}^N(i)\right]$$

$$+\sum_{k=1}^{nnf} C_{34\to12\ T}^{12-34\ N}(i,k)\left[F_{34\to12}^T(k)+FET_{34\to12}^C(i)\right]$$

$$d_{34}^N(i) = -\sum_{k=1}^{nnf} C_{12\to34\ N}^{12-34\ N}(i,k)\left[F_{34\to12}^N(k)+FET_{34\to12}^N(i)\right] \quad [2.22]$$

$$-\sum_{k=1}^{nnf} C_{12\to34\ T}^{12-34\ N}(i,k)\left[F_{34\to12}^T(k)+FET_{34\to12}^C(i)\right]$$

$$+\sum_{k=1}^{nnf} C_{f\to34}^{12-34\ N}(i,k) F_{f\to34}(k)+(Ca-y(i)\alpha)\cos\theta(i)+(Cb+x(i)\alpha)\sin\theta(i)$$

The movement of the rigid body defined by the coefficients a and b and the angle α are introduced to match the position of the shell with the housing one. During this movement, due to the symmetry of the rod and the load in the O x y plane, the effects from the force axial components are not considered.

The equilibrium equations of the shell can be written as:

$$\begin{cases} \sum_{i=1}^{nnf}\left[F_{f\to34}(i)-F_{34\to12}^N(i)-FET_{34\to12}^N(i)\right]\cos\theta(i) \\ \quad +\sum_{i=1}^{nnf}\left[-F_{34\to12}^T(i)-FET_{34\to12}^C(i)\right]\sin\theta(i) = 0 \\ \sum_{i=1}^{nnf}\left[F_{f\to34}(i)-F_{34\to12}^N(i)-FET_{34\to12}^N(i)\right]\sin\theta(i) \\ \quad -\sum_{i=1}^{nnf}\left[-F_{34\to12}^T(i)-FET_{34\to12}^C(i)\right]\cos\theta(i) = 0 \end{cases} \quad [2.23]$$

The forces $F_{f\to34}(i)$ resulting from actions created by the film are known. The rigid body rotation angle α and the tangential components $F_{34\to12}^T(i)$ can be determined by solving the tangential problem. Equations [2.22] and [2.23] constitute a linear system of $nnf+2$ equations with the unknowns being the nnf components $F_{34\to12}^N(i)$ and the coefficients a and b.

2.4.5.2. Tangential problem

The module $\left\|\mathbf{F}^T_{34 \to 12}(i)\right\|$ of the tangential force on node i of the housing–shell interface satisfies the equation:

$$\left\|\mathbf{F}^T_{34 \to 12}(i)\right\| - f\left|F^N_{34 \to 12}(i) + FET^N_{34 \to 12}(i)\right|$$
$$= K_{T_i}\left(d^T_{34}(i) - d^T_{12}(i) - \overline{d^T_{34}(i)} + \overline{d^T_{12}(i)}\right) \quad [2.24]$$

where $d^T_{12}(i)$ and $d^T_{34}(i)$ are the projections in the slippage direction (or in the "slip risk" direction for the stick case) of the tangential movements of point i, which is assumed to belong to the housing or the shells.

The module $\left\|\mathbf{F}^T_{34 \to 12}(i)\right\|$ of the tangential force is expressed using elastothermic forces and those created by the bearing load as:

$$\left\|\mathbf{F}^T_{34 \to 12}(i)\right\| = \sqrt{\left[F^T_{34 \to 12}(i) + FET^C_{34 \to 12}(i)\right]^2 + \left[FET^A_{34 \to 12}(i)\right]^2} \quad [2.25]$$

The circumreferential components $d^C_{12}(i)$ and $d^C_{34}(i)$ of tangential displacements $d^T_{12}(i)$ and $d^T_{34}(i)$ of node i can be written as:

$$d^C_{12}(i) = \sum_{k=1}^{nnf} C^{12-34\ T}_{34 \to 12\ N}(i,k)\left[F^N_{34 \to 12}(k) + FET^N_{34 \to 12}(i)\right]$$
$$+ \sum_{k=1}^{nnf} C^{12-34\ T}_{34 \to 12\ T}(i,k)\left[F^T_{34 \to 12}(k) + FET^C_{34 \to 12}(i)\right]$$
$$d^C_{34}(i) = -\sum_{k=1}^{nnf} C^{12-34\ T}_{12 \to 34\ N}(i,k)\left[F^N_{34 \to 12}(k) + FET^N_{34 \to 12}(i)\right] \quad [2.26]$$
$$- \sum_{k=1}^{nnf} C^{12-34\ T}_{12 \to 34\ T}(i,k)\left[F^T_{34 \to 12}(k) + FET^C_{34 \to 12}(i)\right]$$
$$+ \sum_{k=1}^{nnf} C^{12-34\ T}_{f \to 34}(i,k) F_{f \to 34}(k) - (Ca - y(i)\alpha)\sin\theta(i) + (Cb + x(i)\alpha)\cos\theta(i)$$

Here, the effects from the force axial components are not taken into consideration as well.

The constraints of thermal origin are assumed to be only large enough to create an effect which is only large enough to create a significant impact in the axial direction. The axial components $d_{12}^A(i)$ and $d_{34}^A(i)$ of tangential displacements $d_{12}^T(i)$ and $d_{34}^T(i)$ of node i can be written as:

$$d_{12}^A(i) \simeq d_{ET_{12}^A}(i)$$
$$d_{34}^A(i) \simeq d_{ET_{34}^A}(i)$$

From these movement components, the direction γ of the slippage or the risk of slippage can be determined as:

$$\gamma(i) = \arctan 2\left[\left(d_{34}^C(i) - d_{12}^C(i)\right), \left(d_{34}^A(i) - d_{12}^A(i)\right)\right]$$

and the projections in the slippage direction $d_{12}^T(i)$ and $d_{34}^T(i)$ are given as:

$$d_{12}^T(i) = d_{12}^C(i)\cos\gamma(i) + d_{12}^A(i)\sin\gamma(i)$$
$$d_{34}^T(i) = d_{34}^C(i)\cos\gamma(i) + d_{34}^A(i)\sin\gamma(i)$$

The collinearity between the radial force and the slippage direction can be used in the following expressions:

$$F_{ET_{34\to12}^A}(i) = \left[F_{34\to12}^T(i) + F_{ET_{34\to12}^C}(i)\right]\tan\gamma(i)$$

and:

$$\left\|\mathbf{F}_{34\to12}^T(i)\right\| = \left|F_{34\to12}^T(i) + F_{ET_{34\to12}^C}(i)\right|\sqrt{1+\tan^2\gamma(i)} \qquad [2.27]$$

The shell equilibrium equations can be written as:

$$\begin{cases} \sum_{i=1}^{nnf}\left[F_{f\to34}(i) - F_{34\to12}^N(i) - F_{ET_{34\to12}^N}(i)\right]\cos\theta(i) \\ \quad + \sum_{i=1}^{nnf}\left[-F_{34\to12}^T(i) - F_{ET_{34\to12}^C}(i)\right]\sin\theta(i) = 0 \\ \sum_{i=1}^{nnf}\left[F_{f\to34}(i) - F_{34\to12}^N(i) - F_{ET_{34\to12}^N}(i)\right]\sin\theta(i) \\ \quad - \sum_{i=1}^{nnf}\left[-F_{34\to12}^T(i) - F_{ET_{34\to12}^C}(i)\right]\cos\theta(i) = 0 \\ \sum_{i=1}^{nnf}F_{34\to12}^T(i) + \sum_{i=1}^{nnf}F_{ET_{34\to12}^C}(i) = 0 \end{cases}$$

where the expression [2.27] is used once again:

$$\begin{cases} \sum_{i=1}^{nnf}\left[F_{f\to 34}(i)-F_{34\to 12}^{N}(i)-FET_{34\to 12}^{N}(i)\right]\cos\theta(i)-\sum_{i=1}^{nnf}\frac{\varepsilon(i)\left\|\mathbf{F}_{34\to 12}^{T}(i)\right\|}{\sqrt{1+\tan^{2}\gamma(i)}}\sin\theta(i)=0 \\ \sum_{i=1}^{nnf}\left[F_{f\to 34}(i)-F_{34\to 12}^{N}(i)-FET_{34\to 12}^{N}(i)\right]\sin\theta(i)+\sum_{i=1}^{nnf}\frac{\varepsilon(i)\left\|\mathbf{F}_{34\to 12}^{T}(i)\right\|}{\sqrt{1+\tan^{2}\gamma(i)}}\cos\theta(i)=0 \\ \sum_{i=1}^{nnf}\frac{\varepsilon(i)\left\|\mathbf{F}_{34\to 12}^{T}(i)\right\|}{\sqrt{1+\tan^{2}\gamma(i)}}=0 \end{cases} \quad [2.28]$$

The coefficient $\varepsilon(i)$ worth ± 1 is introduced to take into account the orientation of $F_{34\to 12}^{T}(i)+FET_{34\to 12}^{C}(i)$. The value is assigned at each iteration of the algorithm of the tangential problem based on the analysis of the direction of the slippage or the slippage risk.

Similarly, equations [2.26] can be written as:

$$d_{12}^{C}(i)=\sum_{k=1}^{nnf}C_{34\to 12}^{12-34}{}_{N}^{T}(i,k)\left[F_{34\to 12}^{N}(k)+FET_{34\to 12}^{N}(i)\right]$$

$$+\sum_{k=1}^{nnf}C_{34\to 12}^{12-34}{}_{T}^{T}(i,k)\frac{\varepsilon(i)\left\|\mathbf{F}_{34\to 12}^{T}(i)\right\|}{\sqrt{1+\tan^{2}\gamma(i)}}$$

$$d_{34}^{C}(i)=-\sum_{k=1}^{nnf}C_{12\to 34}^{12-34}{}_{N}^{T}(i,k)\left[F_{34\to 12}^{N}(k)+FET_{34\to 12}^{N}(i)\right] \quad [2.29]$$

$$-\sum_{k=1}^{nnf}C_{12\to 34}^{12-34}{}_{T}^{T}(i,k)\frac{\varepsilon(i)\left\|\mathbf{F}_{34\to 12}^{T}(i)\right\|}{\sqrt{1+\tan^{2}\gamma(i)}}$$

$$+\sum_{k=1}^{nnf}C_{f\to 34}^{12-34}{}^{T}(i,k)F_{f\to 34}(k)-(Ca-y(i)\alpha)\sin\theta(i)+(Cb+x(i)\alpha)\cos\theta(i)$$

The forces $F_{f\to 34}(i)$ resulting from film actions are known. The normal components $F_{34\to 12}^{N}(i)$ were determined upon the solution of the normal problem. Equations [2.24], [2.28] and [2.29] constitute a linear system of $nnf+3$ equations where the unknowns are the nnf components $\left\|\mathbf{F}_{34\to 12}^{T}(i)\right\|$, the coefficients a and b and the angle α.

2.4.5.3. *Contact algorithm*

The algorithm for the resolution of the contact problem at the housing–shell interface is identical to that established for the contact problem in the joint plane (Figure 2.22). Its insertion in the general EHD problem is also similar to the one described for the joint plane (Figure 2.23).

2.5. Case of V engines

In the specific case of V engines, two rod big end bearings are mounted onto the same crank pin. It results in interdependent behavior generated partly due to the elasticity of the common crank pin for the two rods and partly due to the three thrust bearings located between the two rods and between the rods and the side of the crank pin (Figure 2.34). At the same time, both the behavior of the two elastohydrodynamic journal bearings and the three hydrodynamic thrust bearings must be considered. The unknowns of the problem are as follows: (1) the pressure fields in the two rod big end bearings and the three thrust bearings and (2) the eight degrees of freedom of the rods: two eccentricity components (ε_x, ε_y) and two misalignment components (φ_x, φ_y) for each rod. In the first approximation, the axial displacements of the rods are considered to be insignificant and the pressure generated by the thrust bearings is considered too weak to create a significant deformation at the side faces of the two connecting rods (hydrodynamic modeling of the thrust bearings).

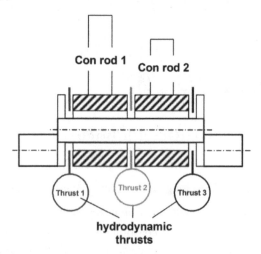

Figure 2.34. *Scheme of the modelization of connecting rod big end bearings for V engines*

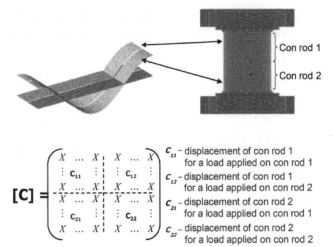

Figure 2.35. *Crank pin compliance*

To determine the pressure field at the rod big end bearing level, the generalized Reynolds equation must be solved, described in detail in Chapter 2 of [BON 14b]. The elasticity of the rods is taken into account through the use of compliance matrices, calculated independently. The compliance of the crank pin ensures the elastic interdependence of the two bearings: a pressure reacting on one of the bearings can modify the deformation at the second. For example, the mesh used to calculate the compliance is shown in Figure 2.35. The general structure of this flexibility matrix is shown in the figure.

The pressure fields for the thrust bearings are calculated by using the same Reynolds equation but in cylindrical coordinates as:

$$\frac{\partial}{\partial r}\left(\frac{rh_b^3}{6\mu}\frac{\partial p}{\partial r}\right) + \frac{\partial}{\partial \theta}\left(\frac{h_b^3}{6r\mu}\frac{\partial p}{\partial z}\right) = V\frac{\partial h_b}{\partial \theta} + 2r\frac{\partial h_b}{\partial t} \qquad [2.30]$$

The expression of equation [4.30] in Cartesian coordinates is given by:

$$\frac{\partial}{\partial x}\left(\frac{h_b^3}{\mu}\left(\frac{\partial p}{\partial x}x + \frac{\partial p}{\partial y}y\right)\right)\frac{x}{r} + \frac{\partial}{\partial y}\left(\frac{h_b^3}{\mu}\left(\frac{\partial p}{\partial x}x + \frac{\partial p}{\partial y}y\right)\right)\frac{y}{r}$$
$$= 6V\left(\frac{\partial h_b}{\partial y}x - \frac{\partial h_b}{\partial x}y\right) + 2r\frac{\partial h_b}{\partial t} \qquad [2.31]$$

This equation can be modified as described in Chapter 2 of [BON 14b] to take into consideration the lubricant filling parameter in order to determine the active and inactive film areas.

The film thickness h_b is given by:

$$h_b(t) = h_{init} + r\left(\tau_{b\_y}\cos\theta + \tau_{b\_x}\sin\theta\right) \qquad [2.32]$$

where h_{init} is the mean thickness and τ_{b_x}, τ_{b_y} are the degrees of freedom related to rotation of the rods $(\varphi_x, \varphi_y)_1$ and $(\varphi_x, \varphi_y)_2$.

The static equilibrium of each rod can be transformed using the following equations:

$$\begin{cases} \mathbf{W}_p - \mathbf{F} = 0 \\ \mathbf{C}_p + \sum \mathbf{C}_{pb} = 0 \end{cases} \qquad [2.33]$$

where $\mathbf{W}_p$ and $\mathbf{C}_p$ represent the force vectors and the moment calculated using the pressure fields in the rod big end bearings and $\mathbf{C}_{pb}$ represents the moment calculated from the pressure fields in the thrust bearings.

The system of equations expressing the EHD behavior of the rod/crank pin combination composed of the following:

– the Reynolds equation governing the hydrodynamic behavior of the lubricating fluid in each of the bearings;

– the Reynolds equation governing the hydrodynamic behavior of the lubricating fluid in each of the stoppers;

– the rod equilibrium equations.

The resolution algorithm is described in detail in Figure 2.36. It is very similar to the EHD algorithm presented in Chapter 5 of [BON 14a]. Figure 2.37 shows the Jacobian matrix of the system, including the variables that appear in each equation.

```
Data initialization
For each time step
    While the domain partition, the thickness, the pressure are not stabilized
        While the domain partition is not stabilized: type 1 problem
            Compute D (modified Reynolds equation)
            Actualize the partition
        End while
        While residual (h,p) > ε (Newton-Raphson method)
            Compute the equation residuals (Reynolds equation and load balance)
            Correction of the pressure for the bearings and the thrusts
            Correction of the rigid displacement parameters
            Compute the elastic deformation
            Modification of journal bearing film thicknesses
            Modification thrust bearing film thicknesses
        End while
    End while
    Write results
End for
```

Figure 2.36. *Resolution algorithm for the case of two connecting rods on the same crank pin*

	p_1	p_2	p_3	p_4	p_5	ε_1	φ_1	ε_2	φ_2
Reynolds con rod 1	X	X	0	0	0	X	X	0	0
Reynolds con rod 2	X	X	0	0	0	0	0	X	X
Reynolds thrust bearing 3	0	0	X	0	0	0	X	0	0
Reynolds thrust bearing 4	0	0	0	X	0	0	X	0	X
Reynolds thrust bearing 5	0	0	0	0	X	0	0	0	X
Force balance con rod 1	X	0	0	0	0	0	0	0	0
Moment balance con rod 1	X	0	X	X	0	0	0	0	0
Force balance con rod 2	0	X	0	0	0	0	0	0	0
Moment balance con rod 2	0	X	0	X	X	0	0	0	0

Figure 2.37. *Organization of the Jacobian matrix system.* p_1, p_2: *pressure fields in crank pin bearings for con rod 1 and 2;* p_3, p_4, p_5: *pressure fields in thrust bearings 1, 2 and 3;* $\varepsilon_i = (\varepsilon_x, \varepsilon_y)$: *eccentricity parameters for con rod i;* $\varphi_i = (\varphi_x, \varphi_y)$: *misalignment parameters for con rod i*

The first modeling example is performed for a linkage system of a Formula 1 engine functioning at 17,000 rpm. Table 2.5 presents the characteristics of the studied system.

Crank shaft radius	20	mm
Connecting rod length	105	mm
Shell width	18.3	mm
Crank pin radius	18	mm
Radial clearance for journal bearings	0.015	mm
Lubricant viscosity	0.012	Pa·s
Feeding pressure	2.5	MPa
Mean clearance for thrust bearings	0.035	mm
V angle	110	°

Table 2.5. *Dimensional data for a V engine at 17,000 rpm*

Figure 2.38 shows the evolution of the maximum pressure values throughout the engine cycle in the two connecting rod big end bearings. The results are compared with those from the classical model where the rods are considered to be independent.

Figure 2.39 shows the same comparison regarding minimum film thickness. Overall, it can be observed that the two connecting rod big end bearings show the same behavior and offset by the angle equal to the angle of the V engine. The comparison with the classical model shows that in this case the interdependence of the two rods is not important and therefore the simple model is sufficient to correctly predict the EHD behavior of the rods.

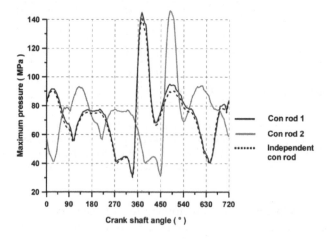

Figure 2.38. *Variation of the maximum pressure for an engine cycle. Case of a V engine at 17,000 rpm*

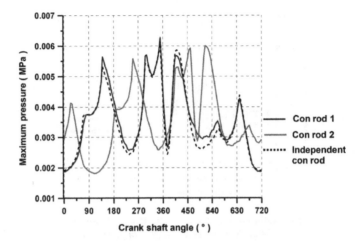

Figure 2.39. *Variation of the minimum film thickness for an engine cycle. Case of a V engine at 17,000 rpm*

Crank shaft radius	40	mm
Connecting rod length	130	mm
Shell width	18	mm
Crank pin radius	24	mm
Radial clearance for journal bearings	0.025	mm
Lubricant viscosity	0.011	Pa·s
Feeding pressure	0.2	MPa
Mean clearance for thrust bearings	0.035	mm
V angle	100	°

Table 2.6. *Data for a V engine at 3,500 rpm*

The second presented example describes an engine functioning at 3,500 rpm. Table 2.6 presents the main characteristics of this setup. Figures 2.40 and 2.41 show how the maximum pressure and the minimum thickness of the film change throughout the engine cycle. As in the previous case, the same bearing is modeled with a classic approach (independent bearings).

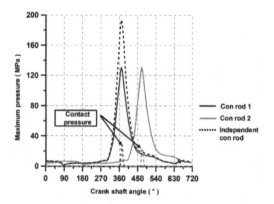

Figure 2.40. *Variation of the maximum pressure for an engine cycle. Case of a V engine at 3,500 rpm*

It is observed that the behavior of the two connecting rod big end bearings is identical and offset by the V engine angle. However, the results obtained using classic calculations are much different, which justifies the use of the model described in this section.

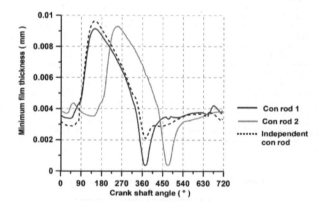

Figure 2.41. *Variation of the minimum film thickness for an engine cycle. Case of V engine at 3,500 rpm*

Figure 2.42 shows the variation of axial force generated in the thrust bearings for an engine cycle for the first configuration studied (an engine working at 17,000 rpm). The axial force is calculated using the pressure fields created in the thrust bearings. The figure shows the pressure field in the three thrust bearings at a crank shaft angle of 470°. It is observed that the maximum pressure is relatively low, which *a posteriori* justifies the modelization without taking into consideration the elastic deformations created by the pressure (hydrodynamic model). However, it can be observed that the axial forces are not balanced and this can result in axial displacement of the connecting rods. Thus, it is possible to conclude that taking into consideration the axial degrees of freedom may improve the generated predictions

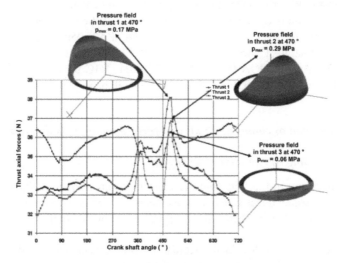

Figure 2.42. *Evolution of the hydrodynamic axial force exerted in the thrust bearings for an engine cycle*

2.6. Examples of connecting rod big end bearing computations

The diversity of engines and the significant number of parameters that may influence the bearing behavior depending on their lubrication do not allow the complete coverage of all configurations. Thus, a few examples of three very different engine types are presented in the following:

– a medium-sized gasoline engine for a large-series car;

– a diesel engine for the same car type;

– an F1 engine.

For the three types of engines, it is shown how the data changes based on the inclusion or exclusion of the following parameters:

– the position of the supply opening on the crank pin;

– the piezoviscosity of the lubricant;

– the non-Newtonian behavior of the lubricant.

For the F1 engine, the connecting rod deformation due to the acceleration field ("intertial deformation") is analyzed. For the gasoline engine, the results generated by the thermo-hydrodynamic models developed in Chapters 2 and 3 are presented.

Other examples of connecting rod big end bearing calculations are presented in different publications by the authors of this book and their collaborators [BON 95, BON 01, DRE 09, FAT 03, FAT 05a, FAT 04, FAT 05b, FAT 05c, FAT 06, FAT 07, FAT 08a, FAT 08b, FAT 08c, FAT 10, FRA 09, FRE 99, GRE 01, GUI 94, HOA 02a, HOA 02b, HOA 03, HOA 05, MIC 04, MIC 07a, MIC 07b, MOR 01, MOR 02, PIF 99, PIF 00, SOU 00, SOU 01a, SOU 01b, SOU 04, TRA 06].

2.6.1. *Presentation of connecting rods and corresponding load diagrams*

2.6.1.1. *Connecting rod for a gasoline engine*

The connecting rod and its finite element mesh are shown in Figure 2.43. The mesh is made using linear tetrahedral elements for the body, cap and screws and prismatic elements for the shells. The materials are indicated in the figure.

Figure 2.44 shows the load diagram of this bearing at 7,000 rpm. At the beginning of the cycle (top dead center), the traction on the rod cap reaches 35 kN. At the bottom dead center, the compression force on the rod exceeds 25 kN.

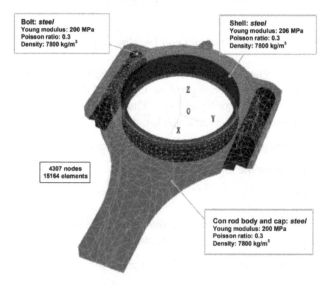

Figure 2.43. *Connecting rod for the gasoline engine; materials and meshes*

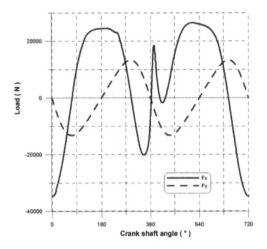

Figure 2.44. *Load diagram for the big end bearing "gasoline" connecting rod at 7,000 rpm*

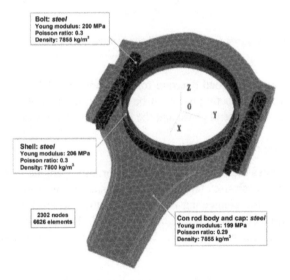

Figure 2.45. *Connecting rod for diesel engine; materials and meshes*

2.6.1.2. *Connecting rod for diesel engine*

The rod and its finite element mesh are shown in Figure 2.45. The elements are the same type as for the connecting rod for the gasoline engine. It must be noted that the diesel engine has a significantly larger connecting rod, especially the shaft and the cap.

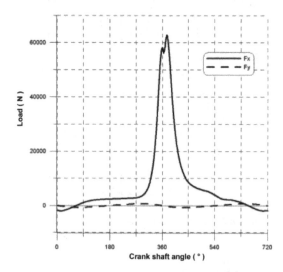

Figure 2.46. *Load diagram for the "diesel" connecting rod big end bearing at 2,000 rpm*

Figure 2.46 shows the load diagram for a bearing at 2,000 rpm. The significant force exerted on the connecting rod must be noted, exceeding 60 kN in a short time period at the compression end phase and the beginning of the combustion phase.

2.6.1.3. *Connecting rod for Formula 1 engine*

Figure 2.47 shows the connecting rod for a Formula 1 engine in the early 2000 s. The finite element mesh uses tetrahedral and prismatic linear elements. To reduce the inertial effects, the connecting rod body and the cap are made from titanium, which is approximately two times lighter than steel. The geometric profile and the dimensions of the connecting rod are very different from the series automobile profiles.

Figure 2.48 shows the load diagram of the bearing at 18,000 rpm. The significant inertial force must be noted, despite the connecting rod and the piston being very light and the reduced crank shaft radius. The force exerted on the rod cap at the top dead center approaches 60 kN. During the second time at the top dead center (engine angle of 360°), the force due to combustion is approximately the same size as the inertial force.

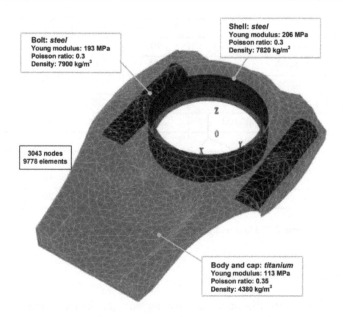

Figure 2.47. *Connecting rod for F1 engine; materials and meshes*

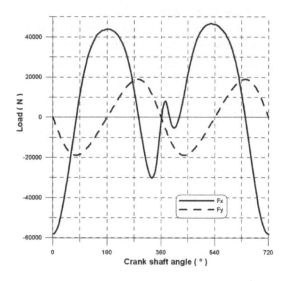

Figure 2.48. *Load diagram for the big end bearing for F1 connecting rod at 18,000 rpm*

Engine kind	Gasoline	Diesel	F1	
Crank shaft radius	46	40	22	mm
Connecting rod length	145	134	106	mm
Connecting rod mass	0.68	0.58	0.30	kg
Con rod mass center position[5]	38	38	40	mm
Moment of inertia for the con rod	2,600	1,800	500	kg·mm^2
Piston and small end shaft mass	0.42	0.61	0.30	kg
Crank pin radius	24	22	18	mm
Shell width	19.5	18	18	mm
Radial clearance	24	22	20	µm

Table 2.7. *Mass and geometry data for linkage and bearings*

2.6.2. *Geometry and lubricant data*

Table 2.7 presents the main dimensions and the inertial parameters for the linkage and bearings for the three engines analyzed. The bearings have a circular profile without lead areas. The crank pins of the "gasoline" and "diesel" bearings are assumed to be perfectly cylindrical. The crank pin of the "F1" bearing is slightly barrel shaped with a 2 µm greater radius in the upper mid-plane when compared to the radius at the side of the bearing. The calculations are carried out isothermally. Table 2.8 presents the lubrication data for the three bearings.

Engine kind	Gasoline	Diesel	F1	
Nominal viscosity (μ_1)	0.005	0.008	0.0045	Pa·s
Piezoviscosity law (p in Pa)	\multicolumn{3}{c}{$\mu = \mu_1 (1 + a p)^b$}	
Coefficient a	\multicolumn{3}{c}{4×10^{-9}}	Pa^{-1}
Coefficient b	\multicolumn{3}{c}{4.6}	
Non-Newtonian Gecim's law ($\dot{\gamma}$ in s^{-1})	\multicolumn{3}{c}{$\mu = \mu_1 \dfrac{K + \mu_2 \dot{\gamma}}{K + \mu_1 \dot{\gamma}}$}	
Coefficient (K)	8,000	22	20	Pa
Viscosity (μ_2)	0.003	0.005	0.0035	Pa·s

Table 2.8. *Lubricant data*

---

5 The center mass position is given with respect to the big end bearing center.

The supply pressure (given at the opening level on the crank pin) is 0.4 MPa for the gasoline engine, 0.3 MPa for the diesel engine and 1 MPa for the F1 engine.

2.6.3. *Analysis of some isothermal results*

The bearing calculation method can provide a large number of numerical results showing:

– changes throughout the cycle for parameters integrated on the bearing surface (for example the torque);

– instantaneous fields for functions defined on the film surface (for example the hydrodynamic pressure fields) or on the surface and inside the surrounding solids (for example the temperature fields);

– fields for functions defined on the film surface, integrated over the duration of the engine cycles (for example the surface wear);

– parameters integrated over the duration of the engine cycles such as the flow rate or the power loss.

A report focusing on the bearing function from a lubrication perspective does not require excessive examination of these results. The focus is primarily on the minimum oil film thickness (MOFT) and on the maximum oil film pressure (MOFP).

These functions are defined as a function of the engine angle over an entire cycle. The first cycle is never used for representational purposes as it is used as the calculation establishment phase (progressive establishment of the actual load, etc.). The second cycle can be used if the parameter values at the end of the cycle are identical to those at the beginning of the cycle. For temperature or wear calculations, even a dozen cycles may be necessary (see Chapters 1 and 5 of [BON 14b]). The instantaneous fields or the fields integrated over the duration of the cycles (film thickness, hydrodynamic pressure, contact pressure, lubricant replenishing, surface wear, etc.) are analyzed in a second time to aid the understanding of the bearing function or the research related to possible malfunctions. The integrated parameters (flow rate and power loss) allow immediate comparisons to be drawn in parametric studies.

For the three connecting rods considered, the changes in MOFT, MOFP and maximum contact pressure, as well as the flow rate and power loss are examined. For these examples, the supply opening position on the crank pin took two values 30.45° and 60° in the engine rotation direction, measured with respect to the most eccentric crank pin generatrix. The calculations were carried out for a piezoviscous

Newtonian lubricant with the dynamic viscosity μ_1 (Table 2.8). To highlight the importance of these characteristics, for each bearing, calculations were carried out in the case of a non-Newtonian fluid (Gecim's law, Table 2.8) and without taking into consideration the piezoviscosity.

Notations

– N PV A60: Newtonian piezoviscous lubricant with the supply positioned at 60°;

– N PV A45: Newtonian piezoviscous lubricant with the supply positioned at 45°;

– N PV A30: Newtonian piezoviscous lubricant with the supply positioned at 30°;

– N IV An: Newtonian isoviscous lubricant with the supply positioned at $n°$;

– NN PV An: Non-Newtonian piezoviscous lubricant with the supply positioned at $n°$.

2.6.3.1. *Minimum film thickness*

The change in MOFT throughout the cycle is very different for the three rods (Figures 2.49–2.51). For a "gasoline" engine because of the high rotational speed and strong centrifugal effect, the minimum thickness is very low throughout almost all of the cycle. Only during the short period at the end of the compression phase and at the beginning of the combustion phase, is there a time during which there is a reversal of the F_x load component (Figure 2.44), such that the surface of the crank pin deviates from the shell so that its center almost matches the bearing center. The supply opening position change does not produce significant changes. As might be expected, not taking into account the piezoviscosity or the thinning effect at high shear rates results in a reduction of MOFT but it remains less than 1 µm.

For a "diesel" engine connecting rod (Figure 2.50), the thickness does not become very low except for the short period corresponding with the beginning of the combustion phase during which the load is very high (Figure 2.46). For the period where the engine angle is between 540° and 640°, it must be noted that the minimum thickness remains very low while the load falls to less than 5% of the maximum load. This is mainly due to the area where the created pressure fields progress onto the area from which the lubricant was expulsed in the previous turn. When the hydrodynamic pressure area encounters an area covered by the supply opening (up to an engine angle of 650°), the minimum thickness rapidly rises. The opening placement at 30° from the reference generatrix on the crank pin creates a lower rise than the placements at 45° or 60°.

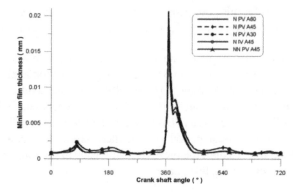

Figure 2.49. *Minimum film thickness for the "gasoline" connecting rod*

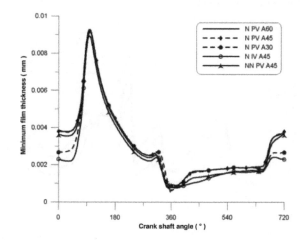

Figure 2.50. *Minimum film thickness for the "diesel" connecting rod*

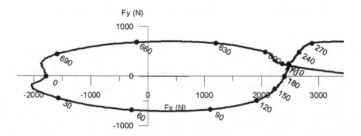

Figure 2.51. *Partial view of the polar load diagram for the "diesel" connecting rod*

One explanation for this poor behavior arises by examining the central part of the bearing polar load diagram (Figure 2.51) between engine angles of 690° and 30°.

The load vector turns approximately 30° while the crank pin turns by 60° with respect to the connecting rod big end bearing. This is a situation analyzed in detail in section 2.6.7, Chapter 2 of Volume 1 [BON 14b]: when the load rotates at half the speed of the shaft, no "oil wedge" effect lift can be generated. Only the squeeze effect ensures a lift: it is necessary that lubricant be present in sufficient quantity at the beginning of this critical period so that there is enough remaining after this period. Not taking into account the piezoviscosity leads to the same effect due to the low values of pressure during this phase.

At an engine angle of about 90°, a rapid rise in the minimum thickness can be seen again: at this angle, the load significantly slows down its rotation, which raises the lift due to the oil wedge effect.

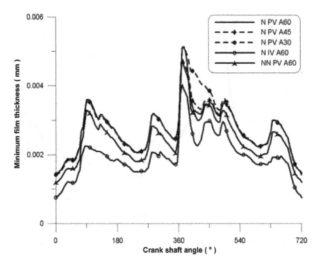

Figure 2.52. *Minimum film thickness for the F1 connecting rod*

The MOFT for F1 connecting rod bearings (Figure 2.52) shows a different profile than the two previous profiles: the high rotational frequency produces an "oil wedge effect", which compensates for the load intensity. The minimum film thickness thus remains approximately the same level over the cycle. The critical period occurs at the top dead center (crank shaft angle 0) in the passage when the load approaches 60 kN and is directed toward the center of the connecting rod cap. Similarly as for the diesel engine connecting rod, the load at this point moves mostly due to inertia and turns at a half the speed of the crank pin. The lift is then solely due to the squeezing effect. Fortunately, this phase is very short as at an engine angle of 30°, the "oil wedge" effect reappears, which results in a rapid rise in the minimum thickness. The position change of the supply opening has a notable effect between engine angles of 360° and 450°, a period during which the minimum thickness is "comfortable". However, this does not mean that the position is not important: *the engine speed varies significantly and verification of good behavior is required at all the speeds to make conclusions about a parameter* (in this case, the position of the opening).

Not taking into consideration the non-Newtonian behavior at high shear rates, as well as the piezoviscosity of the lubricant, leads to reduced minimum thickness values throughout the entire cycle. For this high-speed engine type, these two characteristics definitely need to be taken into account.

2.6.3.2. *Maximum hydrodynamic and contact pressure*

The maximum hydrodynamic pressure for a gasoline engine connecting rod (Figure 2.53) is relatively elevated throughout the majority of the cycle. It does not go below 40 MPa except for a very short period at the final compression phase where the crank pin "floats" in the bearing. The supply opening position does not have an effect on the maximum pressure. However, taking into account the thinning effect and not taking into account the piezoviscosity result in a decrease in pressure.

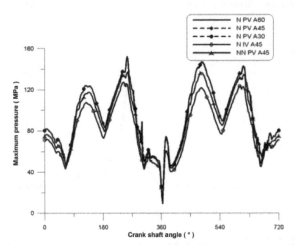

Figure 2.53. *Maximum hydrodynamic pressure for the "gasoline" connecting rod*

The high load supported by the connecting rod big end bearing (perfectly cylindrical) for a gasoline engine leads to the generation of a high contact pressure throughout a significant part of the cycle (Figure 2.54). The levels of contact pressure are strongly influenced by the assumptions used throughout calculations.

Taking into consideration the thinning effect and the piezoviscosity is necessary when carrying out calculations relating to the running-in or wear since they are obtained using contact pressure-based algorithms (Chapter 4 of [BON 14c]).

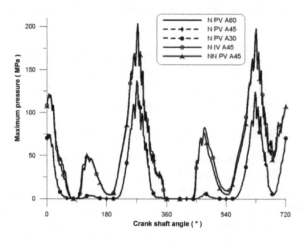

Figure 2.54. *Maximum contact pressure for the "gasoline" connecting rod*

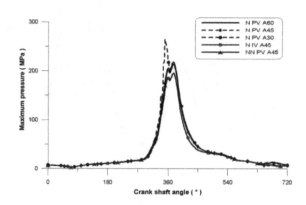

Figure 2.55. *Maximum hydrodynamic pressure for the "diesel" connecting rod*

For a diesel engine connecting rod, the changes in maximum hydrodynamic pressure (Figure 2.55) follow those of the load diagram. The pressure peak is at an engine angle of 350° when the supply opening is located at 30°. The peak is absent in the other setups.

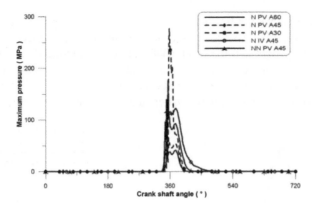

Figure 2.56. *Maximum contact pressure for the "diesel" connecting rod*

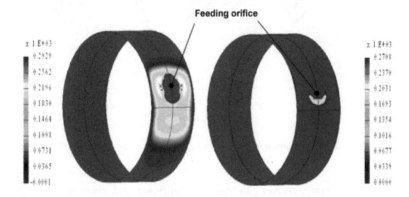

Figure 2.57. *"Diesel" connecting rod. Hydrodynamic and contact pressure fields (MPa) at an angle of 350°*

In terms of the contact pressure (Figure 2.56), a significant increase can be seen due to this supply position. To explain this, the hydrodynamic pressure field needs to be examined at an engine angle of 350° (Figure 2.57, left-hand side).

The supply opening in this case is located in the pressure field. To balance the load, the pressure around the opening must be higher. At the edge of the opening, the hydrodynamic pressure cannot reach the required value as it remains equal to the supply pressure. Thus, the contact pressure must balance the load (Figure 2.57, right side). The contact at the edge of the opening can result in wear. The movement upstream of the supply opening considerably reduces this excessive amount of pressure.

Figure 2.48 shows the bearing load diagram at 18,000 rpm. It can be seen that the inertial forces play a very significant role, despite the connecting rod and the piston being very light and the crank shaft radius being reduced. The force exerted on the connecting rod cap at the top dead center approaches 60 kN. When passing through the top dead center for the second time (engine angle 360°), the force created due to combustion is approximately the same size as the inertial force.

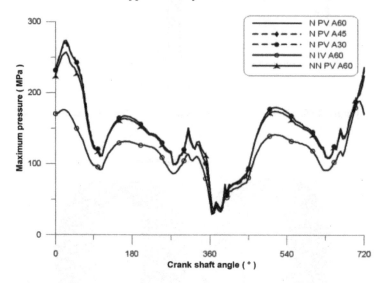

Figure 2.58. *Maximum hydrodynamic pressure for the F1 connecting rod*

For an F1 engine connecting rod, a significant reduction in the maximum pressure can be noted only when the piezoviscosity is not considered (Figure 2.58). The three supply opening positions all give the same maximum pressure values.

2.6.3.3. *Examples of thickness and pressure fields*

For each calculation, the digital processing produces fields illustrating film thickness, hydrodynamic (and sometimes contact) pressure, lubricant filling, shear stress, etc. As examples, Figures 2.59 and 2.60 show the film thickness and pressure fields at an engine angle of 0 (the top dead center at the beginning of the intake phase) for an F1 engine.

Under the inertial force created by the piston, the pressure is exerted exclusively on the connecting rod cap that will "wrap" around the crank pin. The maximum pressure areas are located on the sides of the rod, near the joint plane. It is worth noting that under the pull effect the housing becomes extended on the side of the connecting rod shaft. The maximum thickness reaches 114 µm, which is nearly three times greater compared to the one in the absence of deformation (40 µm).

Figure 2.59. *Film thickness field (µm) at an engine angle of 0 for the F1 connecting rod*

Figure 2.60. *Hydrodynamic pressure field (MPa) at an engine angle of 0 for the F1 connecting rod*

2.6.3.4. *Flow rate and power loss*

The integrated data (flow rate and the power loss) are strongly influenced by the assumptions used in calculations (Table 2.9). For the "gasoline" and F1 engines, the displacement of the supply opening from 60° to 45° and then to 30° allows the increase in flow rate by 30 % and 41 %, respectively. The calculations are carried out at the same viscosity. The power loss is the same for all three positions. In reality, the increased flow rate guarantees better cooling that leads to a higher viscosity and increased power. The assumption of a constant viscosity does not allow this variation. The position of the supply opening seems to have a smaller effect on the diesel engine connecting rod.

Engine	Gasoline		Diesel		F1	
Hypothesis	Flow rate (dm^3/min)	Power (W)	Flow rate (dm^3/min)	Power (W)	Flow rate (dm^3/min)	Power (W)
N PV A60	0.66	849	0.077	70.9	1.26	2,705
N PV A45	0.77	849	0.082	74.8	1.47	2,719
N PV A30	0.88	850	0.088	84.7	1.78	2,744
N IV A60	0.80	719	0.083	72.5	1.29	1,369
NN PV A60	0.86	804	0.084	75.4	1.30	2,427

Table 2.9. *Calculation assumptions related to the flow rate and power loss*

Not taking into consideration the piezoviscosity leads to flow rate values that are 21% higher than for the "gasoline" connecting rod, 8% higher for the "diesel" connecting rod and 2% higher for the F1 rod. The power loss is reduced by 15% for the "gasoline" rod. For the F1 rod, the reduction is very significant from 2,705 to 1,369 W. For this engine type, taking into account the piezoviscosity is definitely a requirement.

The thinning effect due to high shear rates for a non-Newtonian lubricant results in an increase in the flow rate for the three connecting rods (30, 9 and 3%) and the reduction of power by 5% for the "gasoline" connecting rod and 10% for the F1 rod. For the "diesel" rod, a 6% increase in the power loss can be seen but for at a low power level of 75 W.

2.6.3.5. *Effect of accounting the deformation due to inertia*

When the engine speed is high, the deformations due to the acceleration field are important. The comparison between the orbits around the crank pin center obtained with or without taking into consideration these effects (Figure 2.61) for an F1 connecting rod at 18,000 rpm shows significant differences. Under the effect of the acceleration field, the rod is deformed and it is on this deformed housing that the crank pin action is applied. The schematic layout of the connecting rod at different times of the cycle and the layout created taking into consideration the levels of deformation show differences in the profiles of the space between the crank pin surface and the shells (Figure 2.62).

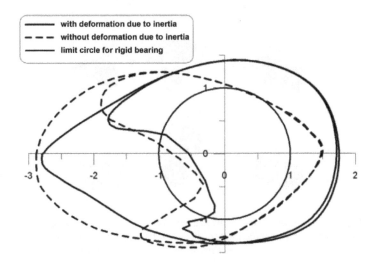

Figure 2.61. *Orbits of the crank pin center (μm) with or without accounting for the deformation due to inertia. F1 connecting rod at 18,000 rpm*

At engine angles of 480° and 540°, the outermost portion of the big end rod is projected outwardly due to the centrifugal action that significantly increases (almost doubles) the film thickness on this side. At an engine angle of 380°, the acceleration reverses: at this moment, the combustion exerts a compression force on the connecting rod and the pressure field (shown by shading in Figure 2.62) located at the side of the rod. The cap is not affected. Under the effect of the acceleration field, the cap shell approaches the surface of the crank pin (top left of diagram) unlike in the case where the inertial deformation is not taken into consideration (bottom left of diagram).

A large difference in film thickness can be seen mainly in the thick film areas. It is interesting to see whether this has a significant impact on the minimum thickness of the film. Figure 2.63 shows the changes in minimum thickness obtained with and without taking into consideration the inertial deformation. The difference is very significant.

Because of the notable differences in film thickness, the integrated values throughout the cycle are also very different: by taking into consideration the inertial deformations, the flow rate is calculated to be 1.26 dm^3/min and the power loss 2,700 W in contrast with 1.39 dm^3/min and 2,844 W.

For the bearings of high-speed engines, taking into consideration the deformations due to acceleration fields is strongly recommended.

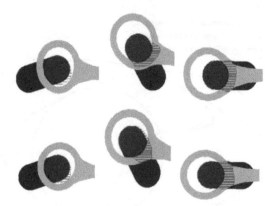

Figure 2.62. *Film thickness and pressure profiles for the F1 connecting rod at 18,000 rpm for engine angles of 380°, 480° and 540° (from left to right). Top: with inertia deformation. Bottom: without inertia deformation*

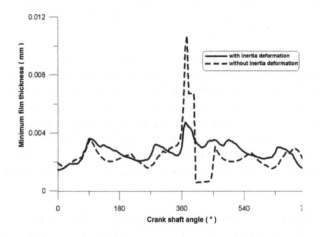

Figure 2.63. *Minimum film thickness with or without accounting for the deformation*

2.6.4. *Influence of mesh downsizing*

A mesh refinement technique is discussed in detail in Chapter 5 of [BON 14a]. The results presented in this section describe the connecting rod of a "diesel" engine, with data given in Tables 2.7 and 2.8 for the two engine speeds of 2,000 and 4,000 rpm. Table 2.10 presents a comparison of some parameters obtained using both a standard model and a model with refined mesh. The integrated parameters, e.g. the flow rate and the power loss, are barely affected by the choice of the model.

	Standard model		Refined model	
Rotational frequency (rpm)	2,000	4,000	2,000	4,000
Maximum hydrodynamic pressure (MPa)	248	212	327	241
Maximum contact pressure (MPa)	19.4	1.50	23.5	10.5
Minimum film thickness (µm)	0.94	1.08	0.99	1.00

Table 2.10. *Influence of mesh downsizing*

At 2,000 rpm, the main difference can be seen in terms of the maximum hydrodynamic pressure. Figure 2.64 shows the final wear fields of the crank pin surface obtained using the two models. Although these fields are similar in appearance, they show significant differences in the supply opening area. On its edges there is a significant wear in the second case and a relatively insignificant wear in the first case. The hydrodynamic pressure fields at an engine contact angle of 384° are shown in Figure 2.65. The decrease in hydrodynamic pressure is compensated by the increase in contact pressure. These two phenomena are described in detail by the refined model; however, the notable size of the elements for the model without refinement leads to a poor representation of the opening shape and a smoothing of the pressure.

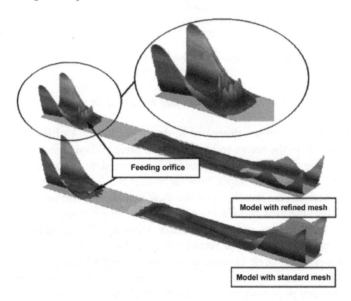

Figure 2.64. *Final wear fields for the crank pin surface at 2,000 rpm*

For an engine speed of 4,000 rpm, the model with the refined mesh predicts a maximum contact pressure at an engine angle of 356°. At this same engine angle, the standard model does not predict pressure contact: the maximum is obtained at 366° with the maximum value being two times lower. Figure 2.66 shows the final wear fields for the shells and the crank pin generated by the two models. The wear predicted using the model without a refined mesh is underestimated for both surfaces.

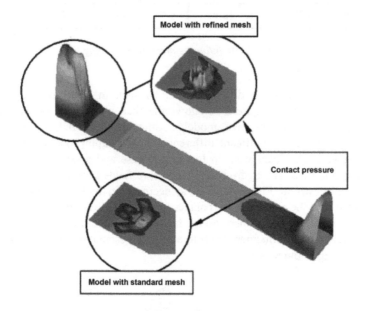

Figure 2.65. *Hydrodynamic and contact pressure fields at 340° (2,000 rpm)*

2.6.5. *Search of potential damage zones due to cavitation*

The cavitation algorithm described in detail in Chapter 3 of [BON 14b] allows the determination of the active and inactive film areas. When an inactive area is seen within an active area, the initial process is assumed to be a "cavitation" type, including the formation of vapor bubbles within the lubricant. As the number of these areas increase, the bubbles increase their volume and merge. If the increase is sufficiently big, the area boundaries can reach the bearing edges or a "separation" area occupied by the air or gas at ambient pressure. The cavitation area is then flooded by gas and becomes a separation area. In contrast, if the pressure rises before the cavitation area is transformed into a separation area, the cavitation bubbles implode. As this occurs over a very short time period, the surrounding fluid returns to the area previously occupied by the bubbles creating sufficient amounts of kinetic energy to cause damage at the solid part surface.

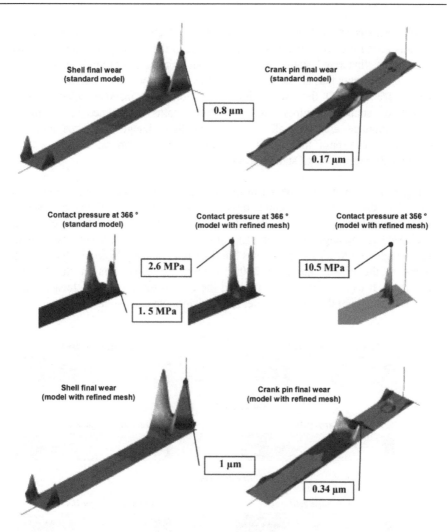

Figure 2.66. *Comparisons for wear and contact pressure fields at 4,000 rpm with respect to the mesh density*

Following the appearance and changes of inactive areas in the film, it is possible to identify those that might pose a significant risk due to damage created through cavitation. To do this, each element of the domain occupied by the film is given a status that can be either "active" or "inactive", and the second status can then be further divided into "cavitation" or "separation" status. At the end of the internal iterative process for each time period, the status of each element is compared with its status in the previous time period. All elements that have gone from an "active" status to an "inactive" status are assigned an initial "cavitation" status. All the "inactive" elements that are on the edge of the bearing or adjacent to a "separation" element are assigned the "separation" condition. An "inactive" element in the "cavitation" status that turns into an "active" status is assumed to belong to a potential damage area. As all the elements that change from "cavitation" to "separation" can pass this novel state onto their neighbors, the algorithm is continuously repeated until stabilization.

Figure 2.67 shows the change in the defined zones based on their status in the case of a "diesel" engine connecting rod at 4,000 rpm. Six fields represent the engine angles between 226° and 264° (intermediate fields were determined). Near a developed bearing angle of 90°, the appearance of an inactive area can be seen at an engine angle of 226°. This area grows but does not reach the bearing edges. It is still present at the engine angle of 260°; however, it disappears in the following calculation. It can be thus considered that the central bearing area is located around the bearing angle of 90° and is a potential cavitation damage area.

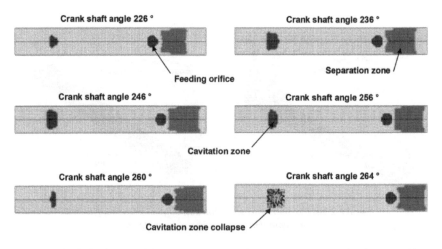

Figure 2.67. *Evolution of the "separation" and "cavitation" zones for the "diesel" connecting rod case at 4,000 rpm*

2.6.6. *Examples taking into consideration thermoelastohydrodynamic effects*

Several thermo-hydrodynamic lubrication models are presented in detail in Chapters 2 and 3 of [BON 14a]. In this section, the "parabolic temperature across the film thickness model" (PTM) and the "mean temperature on the film thickness model" (MTM) is applied to a connecting rod big end bearing for a gasoline engine whose geometrical, material and lubricant characteristics are given in Table 2.11. The corresponding load diagram is shown in Figure 2.68.

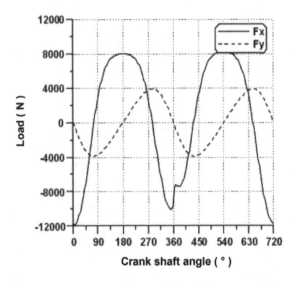

Figure 2.68. *Load diagram for a connecting rod big end bearing at 6,000 rpm*

The viscosity is assumed to follow the law of thermoviscosity (SI units):

$$\mu = 0.0232\, e^{-0.03472\beta(T-70)} + 0.00289$$

where the coefficients are obtained through adjustments based on experimental results.

2.6.6.1. *Model with parabolic temperature profile across the film thickness (PTM)[BON 14a]*

The heat transfer coefficients between the connecting rod and the outside (ambient medium) and between the crank pin and the outside are 50 W/m^2/°C. A large coefficient for the transfer crank pin/supply duct (20,000 W/m^2/°C) is retained in order to set the supply temperature at a quasi-constant value.

Crank shaft radius	40		mm
Connecting rod length	120		mm
Bearing radius	24		mm
Bearing width	18		mm
Radial clearance	30		µm
Rotational frequency	6,000		rpm
	Housing	Crank pin	
Thermal dilatation coefficient	12×10^{-6}	12×10^{-6}	$°C^{-1}$
Thermal conductivity	50	50	$W\,m^{-1}\,°C^{-1}$
Specific heat capacity	500	500	$J\,kg^{-1}\,°C^{-1}$
Density	7,900	7,900	$kg\,m^{-3}$
Young's modulus	200	200	GPa
Poisson ratio	0.3	0.3	
Lubricant density	840		$kg\,m^{-3}$
Lubricant specific heat capacity	2,083		$J\,kg^{-1}\,°C^{-1}$
Lubricant thermal conductivity	0.13		$W\,m^{-1}\,°C^{-1}$
Lubricant viscosity at 70°C	0.0232		Pa·s
Thermoviscosity coefficient	0.03472		$°C^{-1}$
Viscosity limit	0.00289		Pa·s
Feeding pressure	0.5		MPa
Feeding orifice diameter	5		mm
Feeding orifice angular position	30		°
Feeding temperature	80		°C
Ambient temperature	80		°C

Table 2.11. *Big end bearing and lubricant characteristics for a gasoline engine connecting rod*

To approximate the temperature variation in the boundary layer, six Fourier terms are considered.

The connecting rod surface mesh consists of 2,215 triangular elements that exchange with the film and the outside. The crank pin mesh includes on its surface only 828 triangular elements that exchange with the film, the exterior and the supply ducts. All these data lead to a computation time of about 10 h, which are necessary for the thermal convergence of the system that corresponds to 10 engine cycles.

Figure 2.69 shows the variation in minimum film thickness, as well as thermal deformation at the points where the thickness is at its minimum. It can be observed that the surface variation in the thermal field is of the same magnitude as the minimum film thickness. This shows the importance of thermal deformations in the behavior of connecting rod big end bearings.

Figure 2.70 shows the thermal deformations of the housing and the crank pin when engine speed is reached and stabilized. It is possible to see a deformation variation in two directions (radial and axial). This shows that the assumption of a uniform "bulge" for the two solids is not realistic. In addition, if the housing deformation does not vary with the engine angle, the crank pin deformation field is constant in its reference frame but it rotates with the shaft relatively to the housing reference frame.

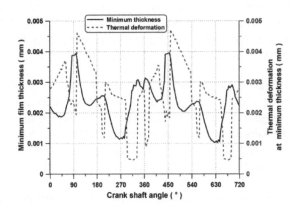

Figure 2.69. *Variation of minimum film thickness and thermal deformation at points where the film thickness is at its minimum*

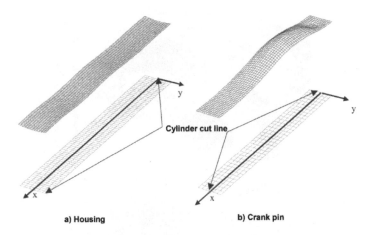

Figure 2.70. *Thermal deformations of the connecting rod housing and the crank pin surface*

The temperature fields that generated these deformations are shown in Figure 2.71. It is possible to observe a significant temperature variation of the housing and the pin at the film/solid interface. For the connecting rod, the temperature varies between 117°C and 127°C and for the crank pin it varies between 85°C and 146°C.

One of the advantages of the PTM model is that it allows the thermal boundary layer to be modeled. Figure 2.72 shows the temperature variation in this boundary layer for a node at the film/housing interface at different rotational angles. Because of a significant temperature variation in time, large temperature gradients can be seen in the layer between 0 and 1 mm in the internal part of the solid. Beyond 1 mm, the temperature does not vary with the crank shaft angle. It is then out of the boundary layer. The figure also shows the thickness of this layer.

The thermal field inside the film in the median symmetry plane at different rotational angles is shown in Figure 2.73. It can be seen that the maximum temperature for engine angles of 0° and 180° is located in the zone where the thickness is at its minimum. For engine angles of 180° and 540°, the maximum temperature can be seen after the minimum thickness zone. Table 2.12 presents the main characteristics calculated for the connecting rod big end bearing.

Although its effectiveness has been proven, the PTM model presented does have a drawback: it requires a significant computing time, especially for studying the connecting rod big end bearings. However, the obtained results provide the coefficients needed for using the MTM model, which is less precise but a lot faster.

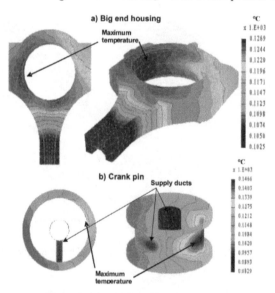

Figure 2.71. *Temperature fields beyond the solid surface boundary layer in stationary engine speed*

Minimum film thickness	1.03	μm
Maximum film pressure	50.5	MPa
Dissipated power	254.5	W
Flow rate	0.84	dm³/min
Shell maximum temperature	127	°C
Crank pin maximum temperature	148	°C

Table 2.12. *Results for a big end connecting rod bearing*

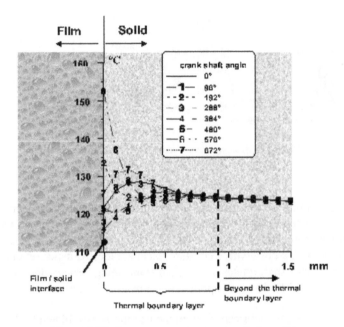

Figure 2.72. *Temperature variation inside the solid at the film/solid interface*

2.6.6.2. Model with mean temperature over the film thickness (MTM) [BON 14a]

This thermal model, which is based on the use of heat transfer coefficients at the film/solid interface, offers the advantage of lower computing times than those necessary for the PTM model. However, the choosing of the film/solid heat transfer coefficient values can prove to be a difficult task.

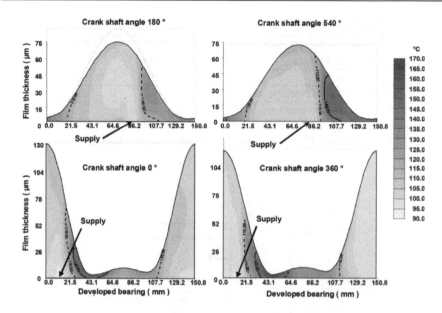

Figure 2.73. *Temperature field in the median symmetry plane for the big end connecting rod bearing*

In the beginning of this section, the mean film/solid heat transfer coefficient values will be determined by comparing the results given by the two models. For this, a simpler problem than the one used for the connecting rod will be examined: it uses a bearing with a bush with a constant thickness located inside an undeformable cylinder, in which a shaft rotates. Fresh fluid is incoming through ducts drilled in the shaft (see Figure 2.74). The load applied on the bearing is constant.

In many studies, the lubricant flow in the bearing is assimilated to the fluid flow between two concentric cylindrical pipes. From this assumption, we can find in the literature numerous equations modeling the fluid/wall heat transfer coefficients using Reynolds, Prandtl and Nusselt numbers. Typically, these numbers are determined experimentally in conditions more or less similar to a bearing. The Nusselt, Reynolds and Prandtl numbers are related via the Huetz equations [HUE 90]:

$$\text{Nu} = c_1 \text{Re}^m \text{Pr}^{0.4} \quad ; \quad \text{Re} = \rho U (2C)/\mu \quad ; \quad \text{Pr} = \mu \, Cp/k \qquad [2.34]$$

where k is the lubricant thermal conductivity and C is the radial clearance of the bearing. c_1 and m are coefficients that depend on geometry and the study conditions.

The heat transfer coefficient H can be written as:

$$H = \text{Nu}\, k / (2C) \qquad [2.35]$$

In [HUE 90], the Nusselt number values can be found in the case of an annular space with isothermal walls depending on the ratio between the outer and inner diameters.

For the bearing defined in Table 2.8 and from the values given by Huetz, it is possible to evaluate the heat transfer coefficient for the film/housing interface at 18,624 W/m^2/°C and between the film and the shaft at 18,713 W/m^2/°C.

The MTM results obtained with these coefficients are compared in Table 2.13 with those of calculations carried out using the PTM model. It is possible to see a significant difference between the two result series. By using the previous model, it is possible to calculate the mean heat flow passing both the film/housing and the film/shaft interface. This can be used to find the heat transfer coefficients, which are approximately 3,600 W/m^2/°C for the shell and 6,000 W/m^2/°C for the shaft. Table 2.13 shows the results obtained with the two models and their heat transfer coefficient values. The observed differences remain below 10% for the main characteristics of the bearing (maximum pressure, minimum thickness, power loss and flow rate). Figure 2.74 shows the thermal fields in the bush and the shaft obtained by using the two models. The geometric position of the maximum temperatures is almost the same in both cases. Yet an underestimation of the mean temperature of the solid can be seen for the MTM model.

	PTM	MTM	
Film/shell heat transfer coefficient (W/m^2/°C)		18,624	3,600
Film/shaft heat transfer coefficient (W/m^2/°C)		18,713	6,000
Minimum film thickness (µm)	7.47	8.15	7.75
Maximum film pressure (MPa)	4.7	4.5	4.5
Dissipated power (W)	159	175	163
Flow rate (dm^3/min)	0.11	0.089	0.102
Maximum shell temperature (°C)	107.4	97.7	99.5
Maximum shaft temperature (°C)	114.4	100.7	102.3

Table 2.13. *Comparisons between models: coefficients derived using bibliography formulas and evaluated using PTM model results*

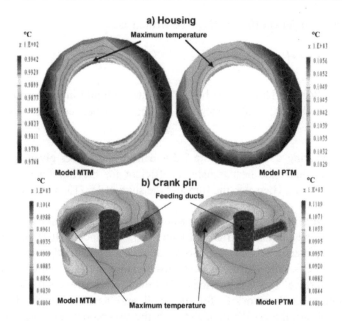

Figure 2.74. *Comparisons between the housing and shaft temperature fields for PTM and MTM models*

For this stationary problem, the computation time needed for convergence is approximately 15 min for the MTM model and 90 min for the PTM model (approximately six times more).

Knowing the heat transfer coefficient for film/bush and film/shaft interfaces and using equations [2.34] and [2.35] with $m = 0.5$ (coefficient typically used in the case of a laminar flow), the coefficient c_{1H} can be deduced for a transfer at the bush and c_{1S} for a transfer at the shaft.

$$c_{1H} = \frac{H_h(2C)}{k\,\mathrm{Re}^{0.5}\,\mathrm{Pr}^{0.33}} = 0.02549 \quad ; \quad c_{1S} = \frac{H_s(2C)}{k\,\mathrm{Re}^{0.5}\,\mathrm{Pr}^{0.33}} = 0.04248 \qquad [2.36]$$

After validating the new coefficients, the film/solid transfers are calculated for different rotational speeds. Table 2.14 shows a comparison between the results obtained using the two models. The differences are insignificant, especially for the low rotational speeds and increase slightly as the rotational speed increases. However, the use of equations [2.34] and [2.35] must be done with caution. To determine the coefficients c_{1H} and c_{1S}, the mean Reynolds number of the bearing was calculated using a mean viscosity of 0.0133 Pa·s, which describes the supply viscosity. In reality, the viscosity significantly varies with temperature.

	3,250 rpm		4,875 rpm		8,125 rpm	
	MTM	PTM	MTM	PTM	MTM	PTM
H_s	2,545		3,117		4,024	
H_h	4,242		5,196		6,708	
h_{min} (μm)	6.31	6.29	7.16	7.02	8.21	7.78
p_{max} (MPa)	4.8	4.9	4.6	4.8	4.5	4.7
Power (W)	65.6	66.6	112.6	112.1	217.9	207.7
Flow rate (dm³/min)	0.060	0.058	0.081	0.082	0.129	0.139

Table 2.14. *Comparison between PTM and MTM models for different rotational frequencies. Stationary case*

The same procedure (equations [2.36]) applied to the connecting rod bearing defined at the beginning of this section allows the film/shell and film/shaft heat transfer coefficients to be estimated. The obtained values are 2,300 and 3,832 W/m²/°C, respectively. Table 2.15 presents the main parameters calculated using both models. Estimating the heat transfer coefficients using the PTM model gives a heat transfer coefficient for the film/shell of about 1,090 W/m²/°C and for the film/shaft a coefficient of about 3,600 W/m²/°C. The obtained results are presented in Table 2.15. The necessary computing times needed for convergence are approximately 80 min for the MTM model and 750 min for the PTM model.

	MTM	MTM	PTM
H_h	2,300	1,090	
H_s	3,832	3,600	
Minimum film thickness (μm)	0.79	0.76	1.03
Maximum film pressure (MPa)	46.9	46.9	50.5
Dissipated power (W)	267.8	261.9	254.5
Flow rate (dm³/min)	0.71	0.70	0.80
Housing maximum temperature (°C)	111.6	109.6	127.0
Shaft maximum temperature (°C)	113.5	115.8	147.8

Table 2.15. *Comparison between PTM and MTM models for a connecting rod big end bearing*

The observed differences between the results obtained from the two models are more pronounced than for a stationary case. However, the main characteristics all display a difference of approximately 10% with the exception of the minimum thickness, which shows a difference of 20%. Figure 2.75 shows the minimum thickness variation with respect to the angle of the crank shaft. Although for certain angles the differences are significant (of about 20%), overall the two curves are almost identical.

All the comparisons examined reveal an underestimation of solid temperature when using the MTM model. However, the main bearing characteristics, e.g. the MOFT, the flow rate or the power loss, display a difference of 10%, which can be satisfactory for a predesign study. The MTM model advantage consumes relatively short computing time when compared with the PTM model.

The use of *a priori* established equations for estimating film/solid heat transfer coefficients must be done carefully. Calculations reveal that these equations give values far from those that lead to results in an acceptable agreement between the two models.

2.6.6.3. *Influence of connecting rod–crank case medium heat transfer coefficients*

The connecting rod big end bearing described in Table 2.8 is considered again. The complexity of the actual connecting rod big end bearing environment makes it very difficult to determine the heat transfer coefficients between the rod and the outside. By varying the coefficients in a range from 20 W/m²/°C (natural convection) to 1,000 W/m²/°C (forced convection), it is possible to quantify the influence of these coefficient values on the bearing behavior.

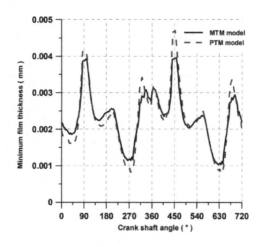

Figure 2.75. *Variation of the minimum film thickness with respect to the crank shaft angle. Comparison between MTM and PTM thermal models*

Figure 2.76(a) reveals the change in maximum temperatures for the housing and the shaft depending on the heat transfer coefficient with the exterior. It is possible to see a logarithmic decline in both the housing temperature (a 17.5% difference between $H = 20$ W/m²/°C and $H = 1,000$ W/m²/°C cases) and the shaft temperature (even though it is less significant for the latter, only a 2.5% difference). Figure 2.76(b) traces the change in differences between solid temperatures with respect to the housing/outside heat transfer coefficient. The difference between the two parameters increases in a logarithmic manner.

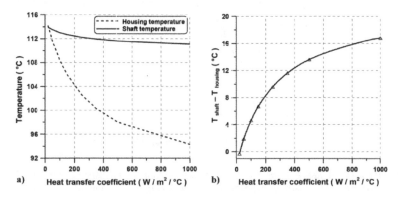

Figure 2.76. *Influence of housing–external medium heat transfer coefficient: a) housing and shaft temperatures and b) difference between housing and shaft temperature*

At the same time, the minimum thickness shows a decrease in variation near approximately 1000 W/m²/°C (Figure 2.77). This variation is directly related to the changes in lubricant viscosity (calculated using the mean film temperature), which also stabilizes when the heat transfer coefficient increases (Figure 2.78).

Figure 2.79 shows the changes in flow rate and power loss. The flow rate shows a decrease in about 30% in the examined heat transfer coefficient range, while the power loss increases by approximately 14%.

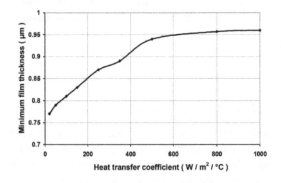

Figure 2.77. *Influence of the housing–external medium heat transfer coefficient on the minimum film thickness*

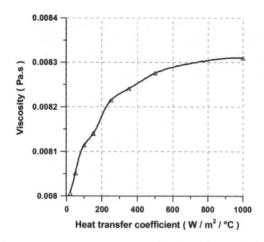

Figure 2.78. *Influence of the housing–external medium heat transfer coefficient on the lubricant viscosity (for the mean film temperature at the end of the cycle)*

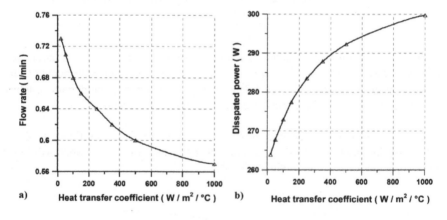

Figure 2.79. *Influence of the housing–external medium heat transfer coefficient on a) the flow rate and b) the dissipated power*

2.6.6.4. *Influence of the shaft–supply duct heat transfer coefficient*

In internal combustion engines, the lubricant circulates under pressure in pipes with a small diameter and smooth wall surfaces. In these conditions, it makes sense to assume that the flow remains laminar regardless of the prevalent operating conditions. It is thus possible to calculate the lubricant/solid heat exchange rates, whose values are between approximately 50 and 100 W/m²/°C [ALE 04]. However, in most numerical models, the temperature of the supply ducts is assumed to be the same as the supply temperature.

An examination of the lubricant/supply duct heat transfer coefficients in the range of between 20 and 20,000 W/m²/°C is enough to reveal the influence that variations in these coefficients have on the main connecting rod big end bearing parameters considered.

Figure 2.80 shows the change in flow rate and power loss depending on the variation of oil/supply duct heat transfer coefficients. An increase in both the power loss and the flow rate can be seen. Cooling of the crank pin leads to an increase in the flow rate, which is the exact opposite of what was seen upon examining the bearing (section 2.6.6.3), where the cooling lead to a decrease in the flow rate.

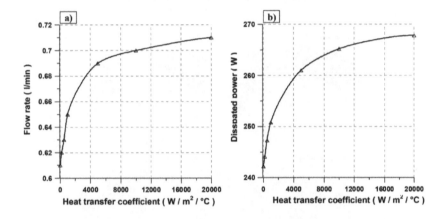

Figure 2.80. *Influence of the shaft–supply duct heat transfer coefficient on the a) flow rate and b) dissipated power*

Figure 2.81 shows a comparison between the thermal fields of the crank pin at different heat transfer coefficient values between the lubricant and the supply ducts. A significant difference (at least 37%) can be noted between the duct temperatures and a coefficient value of 20 W/m²/°C (typical value in literature) and for a coefficient value of 20,000 W/m²/°C, which almost corresponds with the imposed temperature value. This last condition is often used in numerical models. The observed differences in terms of the flow rate and power dissipation (Figure 2.80) can also be seen in maximum pressure values and MOFT (Figure 2.82).

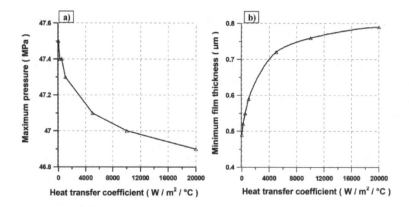

Figure 2.81. *Influence of the shaft–supply duct heat transfer coefficient on the a) maximum pressure and b) minimum film thickness*

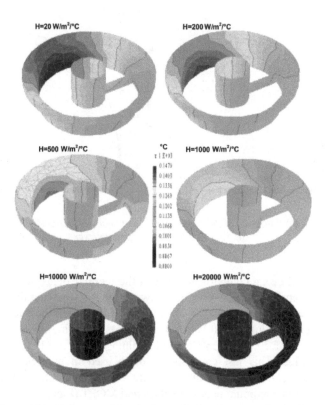

Figure 2.82. *Influence of the shaft–supply duct heat transfer coefficient on the temperature field inside the crank pin*

2.6.6.5. *Procedure for the design of engine bearings*

When comparing the computing times and the precision of the results obtained using the "parabolic temperature model for film thickness" (PTM) and the "mean temperature model for film thickness" (MTM), it is possible to create a procedure for designing bearings under non-stationary load:

– using the MTM model: this model has the advantage that it can estimate bearing behavior fairly well at a reduced computing time;

– using the PTM model: this is necessary for precisely predicting the complete bearing TEHD behavior.

The definition of the housing/exterior and shaft/supply duct heat transfer coefficient values must be considered carefully. It has be seen that by varying the coefficient values from those corresponding to natural convection to those corresponding to forced convection for the walls in contact with the ambient environment or to an imposed temperature for the supply ducts, the main characteristics of the bearing can significantly vary, with the variation reaching up to 50% for the MOFT.

2.6.6.6. *Comparison between EHD, TEHD and non-Newtonian TEHD analysis*

It is assumed that the lubricant follows the non-Newtonian Gecim law:

$$\mu = \mu_1 \frac{K + \mu_2 \dot{\gamma}}{K + \mu_1 \dot{\gamma}}$$

In [GEC 90], Gecim presents the characteristics of a lubricant that he uses to define the non-Newtonian laws at different temperatures. Figure 2.83 shows the variation along with the temperature for two non-Newtonian viscosities μ_1 and μ_2 and the stability coefficient K. Table 2.16 presents the characteristics of both oils. The variation in stability coefficients can be approximated using a second degree polynomial equation:

$$K(T) = 9.2916\ T^2 - 1{,}122.916\ T + 39{,}375$$

	μ_1	μ_2
Thermoviscosity coefficient (°C^{-1})	0.03472	0.03509
Corrective viscosity (Pa·s)	0.00289	0.00163
Viscosity at 70°C (Pa·s)	0.02320 + 0.00289	0.00630 + 0.00163

Table 2.16. *Oil rheological parameters*

Several analyses can be considered as follows:

– TEHD analysis with the thermal PTM model;

– TEHD non-Newtonian analysis with the PTM model;

– EHD analysis with a constant lubricant temperature;

– TEHD analysis with a "global" thermal model, described in Chapter 1 of [BON 14a], with the coefficient α equal to 0.2 (equation [1.1]).

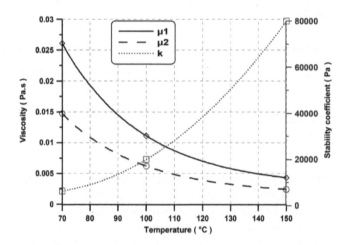

Figure 2.83. *Variations of the viscosity and the stability coefficient depending on temperature*

The case studied considers a connecting rod big end bearing for a gasoline engine whose characteristics are presented in Table 2.10. For EHD analysis, the mean temperature is assumed to be 110°C, which represents the mean lubricant temperature, calculated using the PTM model. For the non-Newtonian analysis, the

Cross law will be used (see equation [1.18] of [BON 14b]) with the two viscosities given in Table 2.16. In a Newtonian case, the μ_1 viscosity is used.

The main bearing operating characteristics are shown in Table 2.17 for each of the four cases studied.

	TEHD PTM	TEHD NN PTM	EHD	"Global" TEHD
Flow rate (dm³/min)	0.84	0.82	0.82	0.56
Dissipated power (W)	254.5	265.5	268.6	407.8
Maximum pressure (MPa)	50.5	49.4	51.5	44.6
Minimum thickness (μm)	1.03	1.00	0.45	1.13

Table 2.17. *Comparison between the different analysis models for the big end bearing of a gasoline engine connecting rod*

A comparison of the results obtained for the cases TEHD PTM and TEHD NN PTM reveals a decrease in the MOFT (3%), the maximum pressure (1.2%) and the flow rate induced by the non-Newtonian effect. At the same time, an increase can be seen in the dissipated power (4.3%). These results show a small decline in the connecting rod performance in the presence of non-Newtonian effects. However, the maximum temperature of the fluid shows a decrease from 175.2°C for the Newtonian case to 162°C for the non-Newtonian case. This represents a 7.5% decrease in temperature. Similarly, the maximum housing temperature slightly decreases from 127°C to 125.2°C while the shaft temperature decreases a little more from 147.8°C to 142.3°C. The non-Newtonian effects thus reduce the performance, but at the same time the film and solid maximum temperatures are a little lower, which may prolong the service life of the bearings.

Figure 2.84 shows the changes in minimum film thickness depending on the crank shaft angle and the orbit of the crank pin center in the housing for each of the four cases. It can be seen that the EHD underestimates the minimum thickness by approximately 0.7 μm.

The TEHD "global" model gives a mean temperature value of about 102°C. With this model, the flow rate is low, which leads to a high power dissipation (almost double, when compared to the other analyses). Significant differences can also be observed in terms of the maximum pressure and minimum thickness.

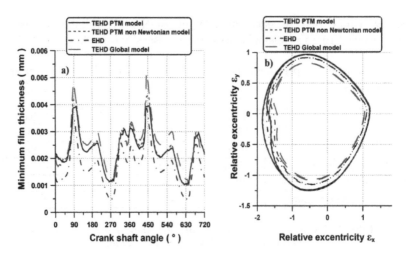

Figure 2.84. *a) Variation of the minimum film thickness. b) Crank pin center orbit*

The comparison between all the models shows the importance of thermal effects on connecting rod big end bearing lubrication analysis and more generally for the bearings under high or non-stationary load. The use of the global thermal model gives very different results from those given by a 3D TEHD model such as the PTM model. However, EHD analysis using the mean temperature given by a 3D TEHD model can be used to predict the bearing behavior with an acceptable degree of confidence.

2.7. Bibliography

[ALE 04] ALEXANDRE A., TOMASELLI L., "Analyse des transferts énergétiques dans les moteurs automobiles", *BM 2900 – 1, Techniques de l'Ingénieur*, 2, Paris, 2004.

[BON 95] BONNEAU D., GUINES D., FRÊNE J., *et al.*, "EHD analysis, including structural inertia effects and a mass-conserving cavitation model", *Journal of Tribology*, vol. 117, pp. 540–547, 1995.

[BON 01] BONNEAU D., HAJJAM M., "Modélisation de la rupture et de la formation des films lubrifiants dans les contacts élastohydrodynamiques", *Revue europeenne des elements finis*, vol. 10, nos. 6–7, pp. 679–704, 2001.

[BON 14a] BONNEAU D., FATU A., SOUCHET D., *Thermo-hydrodynamic Lubrication in Hydrodynamic Bearings*, ISTE, London and John Wiley & Sons, New York, 2014.

[BON 14b] BONNEAU D., FATU A., SOUCHET D., *Hydrodynamic Bearings*, ISTE, London and John Wiley & Sons, New York, 2014.

[BON 14c] BONNEAU D., FATU A., SOUCHET D., *Mixed Lubrication in Hydrodynamic Bearings*, ISTE, London and John Wiley & Sons, New York, 2014.

[BOO 01] BOOKER J.F., BOEDO S., "Finite element analysis of elastic engine bearing lubrication: theory", *Revue européenne des éléments finis*, vol. 10, pp. 705–724, 2001.

[BOO 10] BOOKER J.F., BOEDO S., BONNEAU D., "Conformal elastohydro-dynamic lubrication analysis for engine bearing design: a brief review", *Proceedings of the Institution of Mechanical Engineers, part C*, vol. 224, pp. 2648–2653, 2010.

[DRA 09] DRAGOMIR-FATU R., Etude et modélisation de la lubrification mixte et des modes d'avaries associés dans les paliers moteur, PhD Thesis, University of Poitiers, France, 2009.

[FAT 03] FATU A., HAJJAM M., BONNEAU D., "EHD behavior of non-Newtonian dynamically loaded connecting rod bearings", *Proceedings of the first International Conference on Finite Element for Process, LUXFEM*, pp. 150–162, 2003.

[FAT 04] FATU A., HAJJAM M., BONNEAU D., "Thermo-elastohydrodynamic behaviour of non-Newtonian dynamically loaded journal bearings", *Proceedings of the 12th Conference on EHD Lubrification and Traction*, Romania, 8–9 October 2004.

[FAT 05a] FATU A., Modélisation numérique et expérimentale de la lubrification de paliers de moteur soumis à des conditions sévères de fonctionnement, PhD Thesis, University of Poitiers, France, 2005.

[FAT 05b] FATU A., HAJJAM M., BONNEAU D., "Analysis of non-Newtonian and piezoviscous effects in dynamically loaded connecting-rod bearings", *Journal of Engineering Tribology*, vol. 219, pp. 209–224, 2005.

[FAT 05c] FATU A., HAJJAM M., BONNEAU D., "A fast thermo-elasto-hydrodynamic model for dynamically loaded journal bearing behavior analysis", *Revue européenne des éléments finis*, vol. 14, pp. 237–253, 2005.

[FAT 06] FATU A., HAJJAM M., BONNEAU D., "A new model of thermo-elasto-hydrodynamic lubrication in dynamically loaded journal bearings", *Journal of Tribology*, vol. 128, pp. 85–95, 2006.

[FAT 07] FATU A., BONNEAU D., "EHD lubrication of multi-body common-pin con rod big end bearing system", *STLE/ASME STLE/ASME International Joint Tribology Conference*, San Diego, CA, 22–24 October 2007.

[FAT 08a] FATU A., HAJJAM M., BONNEAU D., "Wall slip in EHD journal bearings", *STLE/ASME Int. Joint Tribology Conf.*, Miami, FL, 20–22 October 2008.

[FAT 08b] FATU A., BONNEAU D., "Wear prediction models for automotive bearings", *7th EDF/LMS Poitiers Workshop*, Poitiers, France, 2 October 2008.

[FAT 08c] FATU A., HAJJAM M., BONNEAU D., "Dynamically loaded low friction bearings based on wall slip boundary condition", *7th EDF/LMS Poitiers Workshop*, Poitiers, France, 2 October 2008.

[FAT 10] FATU A., RICHE I., BONNEAU D., "A finite-element local mesh refinement model for treating mixed-lubrication in con-rodbearings", *International Journal of Surface Science and Engineering*, vol. 4, pp. 155–165, 2010.

[FRA 09] FRANCISCO A., FATU A., BONNEAU D., "Using design of experiments to analyze the connecting rod big-end bearing behavior", *Journal of Tribology*, vol. 131, p. 011101, 2009.

[FRE 99] FRÊNE J., BONNEAU D., "Journal bearing subjected to dynamic loading: the connecting rod-bearing case", *Proceedings of the International Conference on Lubrication Technics*, Istanbul, Turkey, pp. 1–31, 1999.

[GEC 90] GECIM B.A., "Non-Newtonian effects of multigrade oils on journal bearing performance", *STLE Tribology Transaction*, vol. 33, pp. 384–394, 1990.

[GRE 01] GRENTE C., LIGIER J.-L., TOPLOSKY J., *et al.*, "The consequence of performance increases of automotive engines on the modelisation of main and connecting-rod bearings", *Proceedings of the 27th Leeds-Lyon Symposium: Thinning Films and Tribological Interfaces*, Elsevier, pp. 839–850, 2001.

[GUI 94] GUINES D., La lubrification des liaisons compliantes: modélisation et algorithmes, PhD Thesis, University of Poitiers, France, 1994.

[HOA 02a] HOANG L.V., Modélisation expérimentale de la lubrification thermo-élastohydrodynamique des paliers de tête de bielle. Comparaison entre les résultats théoriques et expérimentaux, PhD Thesis, University of Poitiers, France, 2002.

[HOA 02b] HOANG L.V., SOUCHET D., BONNEAU D., "Connecting-rod big end bearing thermo-elastohydrodynamic lubrication (TEHD) – comparison between theory and experiment", *International Journal of Applied Mechanics and Engineering*, vol. 7, pp. 231–236, 2002.

[HOA 03] HOANG L.V., SOUCHET D., BONNEAU D., "Comparison between theory and experiment on behavior of connecting-rod big end bearing under dynamic loading", *11th World Congress in Mechanism and Machine Science*, Tianjin, China, 18–21 August 2003.

[HOA 05] HOANG L.V., SOUCHET D., BONNEAU D., "Thermo-elastohydrodynamic lubrication for connecting-rod big end bearing used in the machine of automobile under dynamic loading", *International Conference on Automotive Technology for Vietnam*, Hanoi, Vietnam, 22–24 October 2005.

[HUE 90] HUETZ J., PETIT J-P., "Notions de transfert thermique par convection", *A 1540 – 1, Technique de l'Ingénieur, traité Génie énergétique*, Paris, 1990.

[MIC 04] MICHAUD P., Modélisation thermo-élastohydrodynamique tridimensionnelle des paliers de moteurs, Mise en place d'un banc d'essais pour paliers sous conditions sévères, PhD Thesis, University of Poitiers, France, 2004.

[MIC 07a] MICHAUD P., FATU A., VILLECHAISE B., "Experimental device for studying real connecting-rod bearings functioning in severe conditions", *Journal of Tribology*, vol. 129, pp. 647–654, 2007.

[MIC 07b] MICHAUD P., SOUCHET D., BONNEAU D., "Thermo hydrodynamic lubrication analysis for a dynamically loaded journal bearing", *Journal of Engineering Tribology*, vol. 221, pp. 49–61, 2007.

[MOR 01] MOREAU H., Mesures des épaisseurs du film d'huile dans les paliers de moteur automobile et comparaisons avec les résultats théoriques, PhD Thesis, University of Poitiers, France, 2001.

[MOR 02] MOREAU H., MASPEYROT P., BONNEAU D., *et al.*, "Comparison between experimental film thickness measurements and elastohydrodynamic analysis in a connecting-rod bearing", *Journal of Engineering Tribology*, vol. 216, pp. 195–208, 2002.

[PIF 99] PIFFETEAU S., Modélisation du comportement thermo-élastohydro-dynamique d'un palier de tête de bielle soumis à un chargement dynamique, PhD Thesis, University of Poitiers, France, 1999.

[PIF 00] PIFFETEAU S., SOUCHET D., BONNEAU D., "Influence of thermal and elastic deformations on connecting-rod big end bearing lubrication under dynamic loading", *Journal of Tribology*, vol. 122, pp. 181–191, 2000.

[SOU 00] SOUCHET D., PIFFETEAU S., BONNEAU D., "Influences des conditions aux limites thermiques sur le comportement d'un palier de tête de bielle", *Proceedings of the 10th Conference on EHD Lubrification and Traction*, VAREHD 10, University of Suceava, Romania, 2000.

[SOU 01a] SOUCHET D., PIFFETEAU S., "Approche par la M.E.F. de la lubrification thermo-élastohydrodynamique des paliers de tête de bielle", *Revue européenne des éléments finis*, vol. 10, pp. 815–847, 2001.

[SOU 01b] SOUCHET D., PIFFETEAU S., BONNEAU D., "Lubrification des paliers de tête de bielle de moteurs thermiques. Modélisation thermohydrodynamique bidimensionnelle", *19th Science and Technology Conference*, Hanoi, Vietnam, 2001.

[SOU 04] SOUCHET D., HOANG L.V., BONNEAU D., "Thermo-elastohydrodynamic lubrication for connecting-rod big end bearing under dynamic loading", *Journal of Engineering Tribology*, vol. 218, pp. 451–464, 2004.

[STE 03] STEFANI F., FEM analysis of the lubrication in connecting rod engine bearings: the influence of structural behaviour on EHD performance, PhD Thesis, University of Genoa, Italy, 2003.

[TRA 06] TRAN T.T.H., Etude expérimentale et modélisation des interactions lubrifiées ou non entre les différents corps d'un palier de tête de bielle, PhD Thesis, University of Poitiers, France, 2006.

[ZIE 00] ZIENKIEWICZ O.C., TAYLOR R.L., *The Finite Element Method*, vol. 2, Solid Mechanics, 5th ed., McGraw Hill, 2000.

3
The Connecting Rod–Piston Link

Compared to studies on the crankshaft–connecting rod link, there are relatively few studies regarding the connecting rod–piston link. Due to its special kinematic arrangement with an alternative rotation movement, this link wastes little energy and rarely causes fatal damage to the engine. However, given the reduction in the size of the engines for a given power, the reliability of this link can no longer be guaranteed with the same degree of certainty. Progress made in calculating the big end connecting rod bearing has led to these techniques being transferred to the bearings in the connecting rod–piston link, with a view to optimize the different elements that constitute it and to guarantee its reliability. The developments discussed in this chapter partially refer to the work of Virgil Optasanu [OPT 00a] and Marco Spuria [SPU 07].

3.1. Geometrical particularities and mechanics of connecting rod–piston link

The connecting rod–piston link connects the two elements with the help of a third element: the axis. Three arrangements are possible:

1) The axis is fitted to the piston and the rotation only takes place between the connecting rod small end and the axis.

2) The axis is fitted to the connecting rod and the rotation takes place between the axis and the piston in the two bearings at the level of the bosses.

3) The axis is in completely free rotation, on both the piston side and the connecting rod side, held in line by elastic rings.

In general, it is accepted that the arrangement is symmetrical with respect to the median plane, in terms of both the force transferred and the shape of the solid parts. In this case, the study can be reduced to the half part of connecting rod small end–axis bearing and one bearing at the axis–piston link.

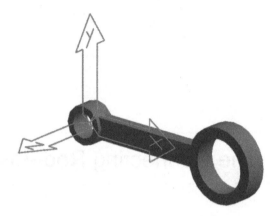

Figure 3.1. *Axis of a connecting rod small end bearing*

Therefore, for arrangements 1 and 2, a single bearing is necessary. For this, it is possible to reuse the models developed for the big end connecting rod bearing. When the axis only turns at the small end of the connecting rod, the latter is taken as a reference (Figure 3.1) and the piston–axis set exerts a given force (the load) on the bearing, thus having the same role as the crankshaft for the big end bearing. This force is given by a load diagram defined for the whole cycle. It can be noted in the examples given in section 1.4.3 that the forces are mainly exerted in the direction of the connecting rod shaft. Combining this with the fact that the relative movement is that of an alternative rotation of the axis $O\mathbf{z}$ with a moderate amplitude (around 17° for an engine in a standard passenger vehicle and 12° for a Formula 1 engine), we are able to deduce that the operating mode of the bearing alternates between phases of oil-film squeezing and fast breaking of this film. The position of the axis in relation to the connecting rod small end is given by eccentricity parameters ε_x and ε_y and possibly, in the event of misalignment, rotation of amplitudes ζ_x and ζ_y (Figure 3.2). When the rotation takes place at the level of the piston bosses, the model stays the same. We should consider the piston as the reference body and the load exerted by the connecting rod and its axis (considered as a single element) on the piston, expressed in relation to the piston.

Although less than the big end connecting rod, the structure of the small end connecting rod is relatively flexible and its deformation under the forces of pressure should be taken into account. However, the deformations resulting from the accelerating fields are usually negligible.

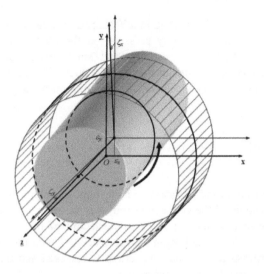

Figure 3.2. *Position parameters of the axis with respect to the connecting rod (case 1) or to the piston (case 2)*

For the third arrangement, the rotational movement of the axis is, in principle, unknown. As it is subject to pressure and shearing actions exerted by both the connecting rod and the piston, its movement can be determined by resolving equations of motion. Nevertheless, this can only be done if combined with the resolution of a set of elastohydrodynamic (EHD) equations that enable the forces to be determined. The corresponding model is described in detail in section 3.4.

Whether on the side of the connecting rod small end or on the piston side, the housing is cast as one (no bolted foot piece) with a cylindrical bush fretted inside. It is even possible for this bush to be left out, and is generally the case on the piston side.

3.2. Lubricant supply

Apart from the small end bearings of connecting rods of very large diesel engines, where a channel drilled in the rod body carries the pressurized lubricant from the big end bearing to the feeding groove, which has been machined into the bush at the small end, the bearings of the connecting rod–piston link are mainly supplied by the lubricant entering from the side through the gaps between the axis and the connecting rod small end or the piston. As far as the small end bearing is concerned, it is supplemented by a channel drilled into its structure, linking a receptacle, close to the end of the connecting rod, to a duct (with a more or less extended median arc, possibly ending in a Y-shaped bifurcation) cut into the surface of the bush. Therefore, the "supply pressure" at the duct is only a result of the acceleration effects that the connecting rod small end undergoes. Therefore, depending on the acceleration, the pressure in the duct will be either higher or lower than the ambient pressure. It is calculated depending on the angle of the engine, θ, by:

$$\Delta p = \rho_f \omega^2 \left(-\cos\theta - \tan\varphi \sin\theta - \frac{R \cos^2\theta}{L \cos^3\varphi} \right) l \cos\varphi \qquad [3.1]$$

where ρ_f is the density of the lubricant, ω is the angular velocity of the crankshaft, φ is the inclination angle of the connecting rod, R is the radius of the crankshaft, L is the length of the connecting rod and l is the length of the channel (Figure 3.3). Relatively low for a slow speed engine, it can reach 0.3 MPa for a Formula 1 engine at 18,000 rpm. At top dead center, the effect is reversed and the direction of the pumping goes from the duct to the receptacle.

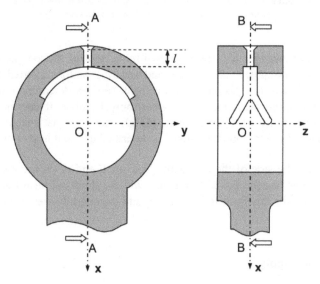

Figure 3.3. *Lubricant supply for a connecting rod small end bearing*

The modeling of thermoelastohydrodynamic (TEHD) of the connecting rod small end bearing by finite elements may use a quadrangular mesh for the film. Although it may be easy to adjust the mesh and the edges of the duct in a circumferential direction, the same cannot be done for the Y-shaped section. Therefore, for this part, if it is used, it is only possible to impose the supply pressure at those points which happen to be inside the boundary of the surface occupied by the duct. The digital form of the duct is, therefore, stepped. This disparity between the digital and true forms of the duct has a small effect on the results.

3.3. Example of computation for a connecting rod small end bearing with the axis embedded into the piston

For this example, we will look at the connecting rod for a medium-sized diesel engine in a standard car. Due to the very low level of dissipation of the bearing at the connecting rod small end, the calculation is carried out in isotherms.

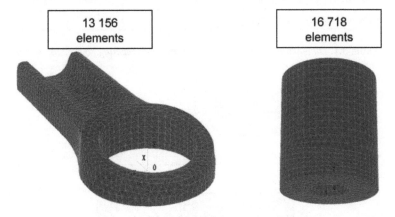

Figure 3.4. *Mesh of the connecting rod small end bearing and its corresponding axis*

The connecting rod and the finite element mesh are shown in Figure 3.4. As the connecting rod essentially supports the force of the compressions, its extremity does not need to have high mechanical resistance. The small end bearing is, therefore, narrower at this side that on the side of the shaft. From a profile view, the small end looks like a viper's head, which is what this type of connecting rod is known as. This variable width of the bearing facilitates lubricant supply.

The mesh is arranged in tetrahedral linear elements. The film mesh is made up of quadrangular quadratic elements (48 elements in the circumferential direction) whose node angles coincide exactly with those of surface nodes for the meshes of the connecting rod small end and the axis. Thanks to the revolution shape of the axis, it is not necessary to apply a rotation operation to the axis' compliance matrix; thus, the nodes of the three meshes will always coincide, which eliminates problems inherent to interpolation (see section 4.4 of [BON 14a]).

Crank shaft radius	45	mm
Connecting rod length	145	mm
External radius of the axis	15	mm
Maximum bearing width	21.5	mm
Minimum bearing width	13.4	mm
Feeding duct length	5	mm
Radial clearance	12	μm
Rotational frequency	2,000	rpm
Lubricant viscosity μ_0 at ambient pressure	0.0035	Pa.s
Piezoviscosity coefficient a	0.0036	MPa^{-1}
Piezoviscosity coefficient b	4.6	

Table 3.1. *Dimensional data for the crank shaft–connecting rod–piston system*

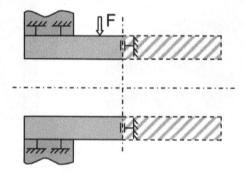

Figure 3.5. *Constraint conditions for the computation of the axis compliance matrix*

To calculate the compliance matrix for the connecting rod small end, it is kept in place by a clamp at the cutting plane. For the axis, it is held in place by a clamp positioned at the piston bosses as indicated in Figure 3.5.

The key characteristics of the bearings as well as the lubricants are given in Table 3.1. The lubricant has piezoviscous properties governed by the power law:

$$\mu = \mu_0 (1 + ap)^b$$

Figure 3.6 shows the components of the load acting on the bearing for a speed of 2,000 rpm. The axial component reaches a value close to 100 kN at the start of the combustion phase. The transverse component does not exceed 180 N.

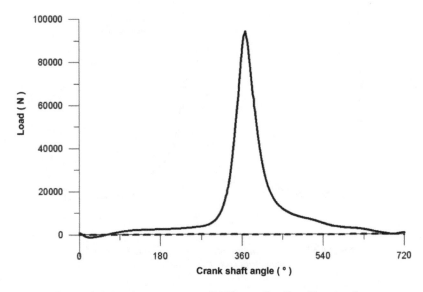

Figure 3.6. *Load components at 2,000 rpm. Small end bearing for a connecting rod high pressure direct Injection (HDI) diesel engine*

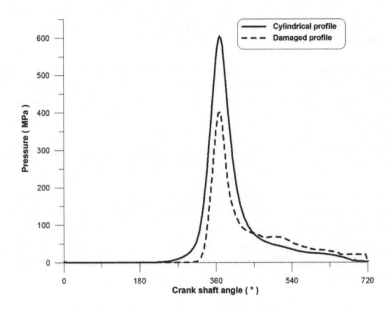

Figure 3.7. *Contact pressure. Small end bearing for a connecting rod HDI diesel engine*

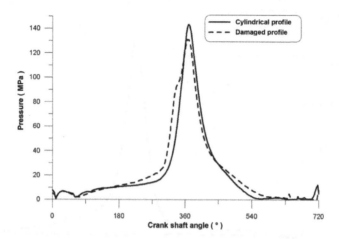

Figure 3.8. *Hydrodynamic pressure. Small end bearing for a connecting rod HDI diesel engine*

Under the effect of the load, the axis bends, which reduces the thickness of the film at the edges of the bearing and creates mixed lubrication occurrence. When the surfaces of the bush and the axis are perfectly cylindrical, the contact pressure reaches very high values (Figure 3.7), around 600 MPa. At the same time, the hydrodynamic pressure remains relatively low (Figure 3.8).

For a new bush and new axis, under the effect of contact pressure, the elasticity threshold of the surface materials is greatly exceeded and a matting effect is produced. After a running-in period, the operating clearance between the surfaces is no longer of a constant thickness. For Figures 3.7 and 3.8, the contact and hydrodynamic pressures have also been traced for a cylindrical shaft and a bush with the profile obtained after a shape adjustment calculation. For this, the algorithm of wear model 1, as described in Chapter 4 of Volume 2 [BON 14b], was used.

After 12 calculation cycles, the contact pressure is significantly lower (Figure 3.7). Since the modification of the average surface profile is coupled with a significant flattening of roughness asperities (see Figure 3.27 of Volume 2), the local contact pressure dips below the elastic limit for materials. After the running-in period, and although the operation of the bearing will continue under mixed lubrication, the profile of the surfaces will be stable.

The thickness of the film (Figure 3.9) reaches its minimum value when the crankshaft angle is around 370°, at the moment when the load is at its maximum. After which, although the load drops dramatically (at an angle of 540°, it is around 5,000 N), the minimum thickness of the film remains very low due to the slow rotational velocity of the axis. It is necessary to wait for the reversal of the load with a crankshaft angle of around 10° during the next cycle to recover a comfortable film thickness. The second rapid rise in the minimum thickness, when the crankshaft angle is around 60°, follows a new load reversal. During these two load reversals, the axis diametrically crosses the bearing which squeezes the film on the opposite side.

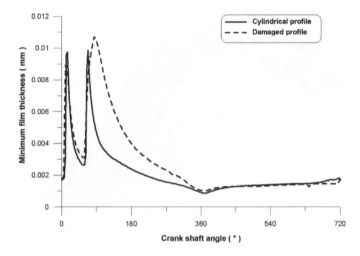

Figure 3.9. *Minimum film thickness. Small end bearing for a connecting rod HDI diesel engine*

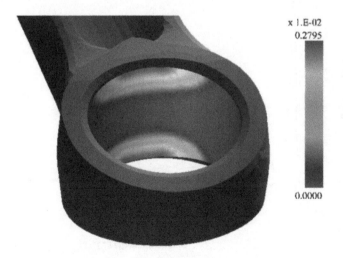

Figure 3.10. *Shape adaptation of the bush. Maximum 2.795 μm. Small end bearing for a connecting rod HDI diesel engine*

Figure 3.10 shows the shape adaptation of the bearing sleeve after 12 calculation cycles. The total pressure (hydrodynamic and contact pressures combined) with the crankshaft angle of 370° is shown in Figure 3.11, superposed with the thickness data. The heavily curved profile in the maximum film thickness zone represents the flexion of the axis. This flexion is at the source of the high pressure areas located at the edges of the bearing in the minimum thickness zones.

Figure 3.11. *Thickness and contact pressure (MPa) at 370° crank shaft angle*

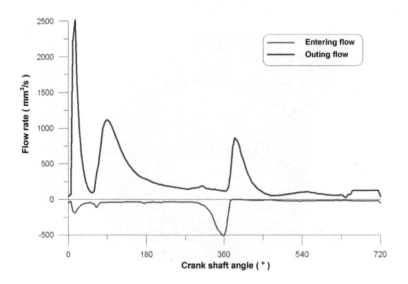

Figure 3.12. *Flows rate exiting and entering through the bearing extremities*

As shown in Figure 3.12, the quantity of lubricant leaving the extremities of the bearing is significantly higher than the amount which enters by the same extremities. Supplementary lubricant is provided by the feeding duct. The suction effect only becomes significant during the compression phase, between crankshaft angles 320° and 360°. By observing the thickness of the film and the pressure at crankshaft angle 350° (Figure 3.13), it is possible to understand why at this moment in the cycle, the pressure in the most significant thickness zone, situated toward the extremity of the connecting rod (the narrowest part of the bearing), is lower than the ambient pressure. The middle part of the bearing is occupied by the groove where the pressure is slightly lower than the ambient pressure (−0.01 MPa) due to the negative acceleration undergone by the connecting rod at top dead center. The area which is not full of lubricant (shown by a dotted line) is gradually filled by the lubricant coming from the groove and the edges of the bearing.

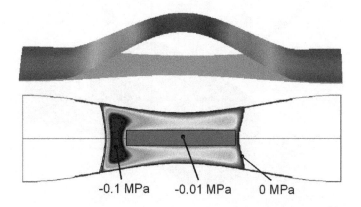

Figure 3.13. *Film thickness and pressure field (part lower than ambient pressure) at 350° crank shaft angle*

3.4. Complete model of the connecting rod–piston link

When the axis of the connection rod–piston link is in free rotation, in relation to the connecting rod and piston, its rotational speed must be determined along with the forces of pressure and shearing that it undergoes. This leads to a definition of a three bearing model, with the axis–small end bearing and the two axis–piston bearings. The latter two can be considered as one single bearing in two unconnected parts. When the overall arrangement has a symmetrical medial plane, only one of the axis–piston bearings and half of the axis–small end bearing is considered.

The following detailed model has been developed by Marco Spuria as part of his thesis work [SPU 05, SPU 07].

3.4.1. *Equations*

Figure 3.14 shows the diagram of the connecting rod small end – piston link. The piston (element 4) can be used as a reference. O z_4 is the housing axis for the axis in the piston bosses. The axis O x_4 is oriented to the piston bottom. The position of the axis (element 3) with respect to the piston is given by the coordinates $\varepsilon_x^{(p)}$ and $\varepsilon_y^{(p)}$, of its center, and the angle χ gives the orientation of one of its radii. The position of the connecting rod (element 2) is given by the coordinates $\varepsilon_x^{(b)}$ and $\varepsilon_y^{(b)}$, from the center of the axis with respect to the center of the small end bearing, and by the angle φ giving the orientation of the connecting rod with regard to the piston.

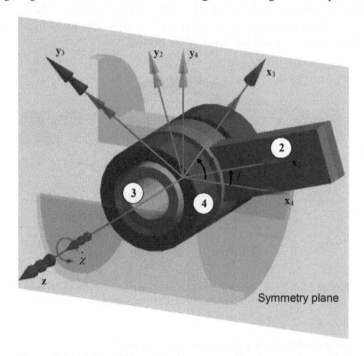

Figure 3.14. *Axis and kinematical parameters for the complete model*

The laws of dynamics applied to the axis are as follows (notations defined in Chapter 1):

$$\begin{cases} m_3 \ddot{x} = X_{43} + X_{23} \\ 0 = Y_{43} + Y_{23} \\ I_3 \ddot{\chi} = M_{43} + M_{23} \end{cases} \quad [3.2]$$

The first two equations are fully determined: the acceleration $\ddot{x}$ of the piston is calculated (not taking the mobility of the links into account) depending on the engine angle θ. The components of the force X_{23}, Y_{23}, X_{43} and Y_{43}, expressed on the basis of $\mathbf{x}_4$, $\mathbf{y}_4$ connected to the piston, are, therefore, entirely determined from the pressure of the combustion chamber (see Chapter 1).

The third equation connects the angular acceleration of the axis $\ddot{\chi}$ to the moments it undergoes. I_3 represents the moment of inertia of the piston axis in relation to its revolution axis, and M_{23} and M_{43} are the friction torques exerted on both the piston and the connecting rod. These friction torques are calculated by integrating the shearing stresses for each of the link bearings. The shearing stress is obtained from the fields of hydrodynamic pressure and possibly contact pressure. These fields of pressure are determined by resolving Reynolds and contact equations. The Reynolds equation uses the relative velocity of the surfaces, expressed directly as a function of $\dot{\chi}$. The system of integrodifferential equations made up of the equation of moments and the Reynolds equations for the two[1] bearings, is completed by the two equilibrium equations for the load borne by each bearing.

This system is split into two subsystems:

– subsystem 1: the four load equilibrium equations (two per bearing) and the two discretized Reynolds equations;

– subsystem 2: dynamic equation on moments.

At each time step, the successive resolution of these two subsystems should be repeated until solutions are stabilized.

The unknown fields of subsystem 1 are the two pressures $\mathbf{p}^{(b)}$ and $\mathbf{p}^{(p)}$ for each of the two bearings (marked as $^{(b)}$ for the axis–connecting rod bearing and $^{(p)}$ for the axis–piston bearing) and the four eccentricity parameters $\varepsilon_x^{(b)}$, $\varepsilon_y^{(b)}$, $\varepsilon_x^{(p)}$ and $\varepsilon_y^{(p)}$, give the relative position of the axis with respect to the connecting rod on one side and to the piston on the other. These unknown values are noted in the following vectorial form:

$$\mathbf{x} = \left\{ p_1^{(b)} \ldots p_{nnb}^{(b)} \; p_1^{(p)} \ldots p_{nnp}^{(p)} \; \varepsilon_x^{(b)} \; \varepsilon_y^{(b)} \; \varepsilon_x^{(p)} \; \varepsilon_y^{(p)} \right\}^T \qquad [3.3]$$

nnb and *nnp* are, respectively, the number of nodes of the film mesh of each bearing.

---

[1] Following this, the two axis–piston bearings will be considered as one bearing. This will also be the case for all of their characteristics; a single field of pressure (in two separate parts), a single moment, etc.

The Reynolds equation uses vectors $\mathbf{h}^{(b)}$ and $\mathbf{h}^{(p)}$ of the film thicknesses at the mesh nodes as intermediate parameters. These thicknesses include the usual surface deformation, represented by vectors $\mathbf{d}^{(a)}$, $\mathbf{d}^{(b)}$ and $\mathbf{d}^{(p)}$. Vector $\mathbf{d}^{(a)}$ of the axis surface deformation is made up of two subvectors $\mathbf{d}^{(ab)}$ and $\mathbf{d}^{(ap)}$ of the respective dimensions of *nnb* and *nnp* which correspond to the axis–connecting rod and axis–piston bearings.

The relations between the displacement and pressure vectors are written as follows:

$$\mathbf{d}^{(ab)} = [C^{(ab)}]\, \mathbf{p}^{(b)}$$

$$\mathbf{d}^{(ap)} = [C^{(ap)}]\, \mathbf{p}^{(p)}$$

$$\mathbf{d}^{(b)} = [C^{(b)}]\, \mathbf{p}^{(b)} \quad\quad [3.4]$$

$$\mathbf{d}^{(p)} = [C^{(p)}]\, \mathbf{p}^{(p)}$$

where the compliance matrices are determined by using the procedure described in Chapter 4 of [BON 14a]. For reasons of simplicity, these relations, and therefore the pressure fields, are expressed relative to the piston or the connecting rod small end depending on the bearing in question. The residuals of the equilibrium equations of the load exerted on the axis, projected in relation to x_4, y_4 of the piston, are written as follows:

$$R\left(W_x^{(p)}\right) = X_{43} + \iint_{\Omega^{(p)}} p^{(p)}(x,z) \sin\theta(x,z)\, dxdz$$

$$R\left(W_y^{(p)}\right) = Y_{43} + \iint_{\Omega^{(p)}} p^{(p)}(x,z) \cos\theta(x,z)\, dxdz$$

and:

$$R\left(W_x^{(b)}\right) = X_{23} + \iint_{\Omega^{(b)}} p^{(b)}(x,z) \cos(\theta(x,z)-\varphi)\, dxdz$$

$$R\left(W_y^{(b)}\right) = Y_{23} + \iint_{\Omega^{(b)}} p^{(b)}(x,z) \sin(\theta(x,z)-\varphi)\, dxdz$$

where $\Omega^{(p)}$ and $\Omega^{(b)}$ are, respectively, the areas occupied by the film in the axis–piston and axis–connecting rod bearings.

The nonlinearity of subsystem 1 means that it is necessary to use a Newton–Raphson algorithm (Chapter 4 of [BON 14a]). The respective notations will be $E^{(b)}$ and $E^{(p)}$ for the Reynolds equations regarding the axis–connecting rod bearing on one side and the axis–piston bearing on the other. $E^{(Wbx)}$, $E^{(Wby)}$, $E^{(Wpx)}$ and $E^{(Wpy)}$ will be used for the four load equilibrium equations, for each bearing and each component.

If $\mathcal{E}$ represents the entirety of the above equations, the system of EHD equations:

$$\mathcal{E}(\mathbf{x}) = 0 \qquad [3.5]$$

gives rise to the following Jacobian matrix:

$$J = \begin{bmatrix} \left[\dfrac{\partial E^{(b)}}{\partial p^{(b)}}\right] & \left[\dfrac{\partial E^{(b)}}{\partial p^{(p)}}\right] & \left[\dfrac{\partial E^{(b)}}{\partial \varepsilon_x^{(b)}}\right] & \left[\dfrac{\partial E^{(b)}}{\partial \varepsilon_y^{(b)}}\right] & [0] & [0] \\ \left[\dfrac{\partial E^{(p)}}{\partial p^{(b)}}\right] & \left[\dfrac{\partial E^{(p)}}{\partial p^{(p)}}\right] & [0] & [0] & \left[\dfrac{\partial E^{(p)}}{\partial \varepsilon_x^{(p)}}\right] & \left[\dfrac{\partial E^{(p)}}{\partial \varepsilon_y^{(p)}}\right] \\ \left[\dfrac{\partial E^{(Wbx)}}{\partial p^{(b)}}\right] & [0] & [0] & [0] & [0] & [0] \\ \left[\dfrac{\partial E^{(Wby)}}{\partial p^{(b)}}\right] & [0] & [0] & [0] & [0] & [0] \\ [0] & \left[\dfrac{\partial E^{(Wpx)}}{\partial p^{(p)}}\right] & [0] & [0] & [0] & [0] \\ [0] & \left[\dfrac{\partial E^{(Wpy)}}{\partial p^{(p)}}\right] & [0] & [0] & [0] & [0] \end{bmatrix} \qquad [3.6]$$

We may note that the Reynolds equations for the two bearings have been combined due to the interdependence of the elastic deformations: the thickness h, therefore, the surface deformation, is a parameter in the Reynolds equation, and consequently the axis surface deformation depends on the values of the two pressure fields (equation [3.3]).

3.4.2. *Integration of dynamics equation*

The dynamics equation applied to the axis (third equation [3.2]) is integrated using the following SS22 algorithm (*Single Step*, second-order of approximation of the function, second-order of the equation) as proposed in [ZIE 00]. This algorithm approximates, in the interval of the integration step, the function required by a second degree polynomial. The second degree polynomial must satisfy the standard dynamics equation, averaged over the interval: the error within the integration interval Δt, weighted by a judiciously chosen function $W(t)$ must be negligible.

The equation to be integrated is in the following form:

$$I_3 \ddot{\chi} + C = 0 \qquad [3.7]$$

where C represents the opposite of the sum of the friction torque being exerted on the axis. We should consider a given moment t in the interval $[t_n, t_{n+1}]$. The approximation of the function χ is given by its development of a Taylor series, limited to second order:

$$\tilde{\chi}(t) = \chi_n + \dot{\chi}_n (t - t_n) + \frac{1}{2} \ddot{\chi}_n (t - t_n)^2 \qquad [3.8]$$

where $\dot{\chi}_n$ and $\ddot{\chi}_n$, respectively, represent the values of the first and second derivatives of χ at time t_n.

The weighted residual, averaged over the interval $[t_n, t_{n+1}]$ with a duration Δt, of the dynamics equation is written as follows:

$$R = \int_0^{\Delta t} W(t) \left(I_3 \ddot{\chi} + C \right) dt$$

The approximation [3.8] requires the second derivate of χ to be constant over the interval $[t_n, t_{n+1}]$. The cancelling out of the weighted residual then leads to the following relation:

$$\ddot{\chi}_n = -\frac{\int_0^{\Delta t} W(t) C \, dt}{I_3 \int_0^{\Delta t} W(t) \, dt}$$

If it is supposed that the evolution of the couple C in the interval $[t_n, t_{n+1}]$ is linear, it follows that:

$$\ddot{\chi}_n = -\frac{1}{I_3} \left(C_n + \frac{(C_{n+1} - C_n)}{\Delta t} \frac{\int_0^{\Delta t} W(t) t \, dt}{\int_0^{\Delta t} W(t) \, dt} \right)$$

The choice of a constant weighting function gives the following:

$$\ddot{\chi}_n = -\frac{1}{I_3} \frac{C_n + C_{n+1}}{2} \qquad [3.9]$$

Equation [3.8] enables the following:

$$\chi_{n+1} = \chi_n + \dot{\chi}_n \Delta t + \frac{1}{2}\ddot{\chi}_n \Delta t^2 \qquad [3.10]$$

$$\dot{\chi}_{n+1} = \dot{\chi}_n + \ddot{\chi}_n \Delta t \qquad [3.11]$$

The general algorithm is given in Figure 5.15.

> While $\dot{\chi}_{n+1}$ and $\mathbf{p}_{n+1}$ are not stabilized
>
>   Compute $\ddot{\chi}_{n+1}$, equation (3.9)
>
>   Compute $\dot{\chi}_{n+1}$, equation (3.11)
>
>   Compute χ_{n+1}, equation (3.10)
>
>   Compute $\mathbf{p}_{n+1}$, EHD equations (3.5) with the new angular velocity $\dot{\chi}_{n+1}$
>
>   Compute C_{n+1} from the new pressure field $\mathbf{p}_{n+1}$
>
> End while

Figure 3.15. *Algorithm for computation of axis rotational velocity*

During the repeated resolution of the system {EHD equations – equation [3.5]}, over the interval [t_n, t_{n+1}], the value of C_{n+1} in the relation [3.8] is gradually improved.

3.4.3. *Piston structural model*

The piston essentially undergoes the effect of gas pressure on the upper part, the effects of contact with the segments or with the cylinder liner at the level of its skirt and the effect of the connecting rod via the axis. The forces of gas pressure in the area between the liner and the piston, whether above the fire-stop segment or between the segments, play a less significant role in the equilibrium of the piston since they have a negligible resultant effect compared to other forces. The result of the combination of these forces balances the dynamic resultant, product of mass via the acceleration of the center of mass. The resultant moment of the forces balances the dynamic moment. This second vectorial equation governs the low amplitude rotation movements that the piston may have. These rotation movements are not taken into account.

In order to determine the structural deformation of the piston under the effect of these mechanical forces, a finite element model is used, in which the forces acting (including the inertial force considered as a volume force factor), are transformed into discrete forces acting on each of the nodes of the structure. Following a process described in Chapter 4 of [BON 14a], the deformation is obtained by an algebraic calculation using a compliance matrix and force or nodal pressure vectors. To obtain the compliance matrix, unitary forces are applied successively to each of the "useful" nodes and a deformation calculation is carried out using the finite element model.

For a slender structure, such as a connecting rod, the area where the deformations are surveyed – the surface of the bearing – is far from the areas where the forces act apart from those acting on the surface of the bearing itself. Therefore, a model corresponding to a truncated part of the connecting rod (e.g. see Figure 3.4) can be produced and the cut plane can be used as a reference surface for the calculation of deformations (clamped section). For a diesel engine piston, which is relatively sturdy, it can be assumed that the upper surface of the piston cannot be deformed and can be used as the reference surface. This technique cannot be used for a Formula 1 engine which, due to its shape and the intensity of the forces it undergoes, will experience deformations around the whole of its structure.

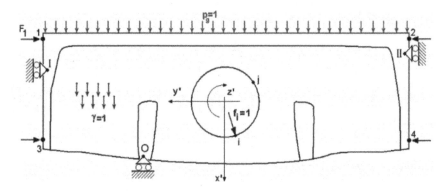

Figure 3.16. *Mechanical actions applied to the piston*

For the model subsequently developed, it can be assumed in any case that, due to the low intensity of the contact forces on the piston segments – cylinder liner and piston – compared to the other forces, these can be replaced by four concentrated forces applied to four nodes whose positions are given in Figure 3.16 (nodes 1, 2, 3 and 4). In the figure, the friction component (oriented in relation to **x**) for these forces is negligible, but can be taken into account.

The combination of the "elementary solutions", made up of radial displacements at nodes j of the surface of the axis–piston bearing, correspond to the following single actions:

– pressure of gases $p_g = 1$;

– acceleration of the piston $\gamma = \ddot{x} = 1$;

– normal piston–cylinder contact force at each of the nodes 1, 2, 3 and 4;

– normal force at each of the nodes i of the surface mesh of the axis–piston bearing.

These produce, once implemented, the required compliance matrix.

Due to the symmetry of the structure and load, a half-model in the direction z is used with a null normal displacement condition in the cut plane at $z = 0$. For the finite element calculation, the piston must be maintained by positioning constraints which only prevent the solid body displacement in directions x and y. These are shown in Figure 3.16 as O, I and II. The positions of x for constraints I and II must be different.

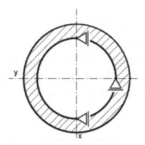

Figure 3.17. *Isostatic constraints for the axis*

Each load applied during the calculation of elementary solutions gives rise to reactive forces at each maintenance point O, I and II and, therefore, interference deformations in the vicinity of these points. However, during the combination of elementary solutions at the moment of the EHD calculation, the forces and the resulting deformations will disappear since the linear combination of the actions taken into account as a whole include all non-negligible forces, but also the "inertial forces" and, therefore, will produce a negligible resultant.

To calculate the deformation of the axis, there is a problem similar to that encountered with the piston. A half-model with null displacement conditions z in the plane $z = 0$ is used. The choice of isostatic equilibrium conditions is more delicate than for the piston. The points chosen will necessarily be closer to the surface of the bearings (Figure 3.17). In the event of poor equilibrium of the combined forces exerted on the axis, the residual reactions may lead to modifications in the thickness profile of the film at the level of these points. If this is the case, it is advisable to choose other points, if possible situated in the areas where the thickness of the film is never very low, since flaws regarding the thickness of the film in these areas would have a reduced impact.

3.4.4. *Example: the piston–axis–connecting rod small end link for a Formula 1 engine*

In his thesis work, Marco Spuria [SPU 07] studied the piston–axis link in a Formula 1 engine in detail. The results discussed below are extracts of this work. They concern two engine speeds: first with a very high rotation speed (20,000 rpm) but with almost negligible engine torque, for which the effects due to inertia are dominant, and a lower speed engine (17,000 rpm) with a maximum amount of engine torque.

3.4.4.1. *Data for engine, bearings and lubricant*

Table 3.2 shows the main characteristics of the connecting rods, bearings and lubricant. It can be assumed that the lubricant has piezoviscous properties in accordance with the power law. The viscosity at ambient pressure is given for the operating temperature of around 200°C. The entirety can be considered in isotherm.

Crank shaft radius	22	mm
Connecting rod length	106	mm
External radius of the axis	12	mm
Small end bearing width	20	mm
Piston bearing width	12	mm
Radial clearance for the small end bearing	12	µm
Radial clearance for the piston bearings	22	µm
Lubricant viscosity μ_0 at ambient pressure	0.0035	Pa.s
Piezoviscosity coefficient a	0.0036	MPa^{-1}
Piezoviscosity coefficient b	4.6	

Table 3.2. *Data for engine, bearings and lubricant*

3.4.4.2. *Characteristics for materials and meshes*

Table 3.3 shows the main characteristics of the solid materials.

The meshes of the piston, axis and connecting rod are represented by Figures 3.18 and 3.19. In order to be able to use a half-model, these meshes must have a symmetrical plane, the equation for which is $z = 0$. The elements are linear tetrahedrons with four nodes, except for the bush which is meshed in linear prismatic elements with six nodes.

	Material	E (GPa)	ν	ρ (kg m^{-3})
Piston	Aluminum	70	0.33	2,700
Axis	Steel	210	0.3	7,800
Ring	Steel	210	0.3	7,800
Connecting rod	Titanium	110	0.33	4,650

Table 3.3. *Data for the materials of piston–connecting rod link*

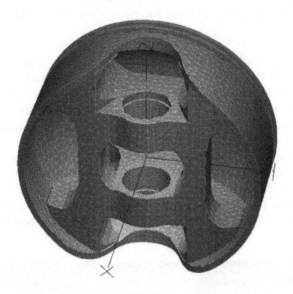

Figure 3.18. *Piston mesh*

Figure 3.19. *Meshes for the axis and the connecting rod small end*

3.4.4.3. *Running parameters*

For axis–piston bearings, the lubricant is supplied at the sides of the bearings. For the small end bearing, it is supplemented by an opening placed on the bush linked by a duct to the big end bearing of the connecting rod. The outlet pressure of this duct is predominantly dictated by the dynamic effect due to acceleration (relation [3.1]).

The value of the radial clearance in each of the bearings used for the calculation is estimated with the nominal radial clearance and the effect of the differential expansions of the various materials: $13 \times 10^{-6}\,°C^{-1}$ for steel, $23 \times 10^{-6}\,°C^{-1}$ for aluminum and $10 \times 10^{-6}\,°C^{-1}$ for titanium. With heating, the radial clearance increases for piston bearings and reduces for the connecting rod bearing.

The deformation of the piston, the axis and the connecting rod is initially calculated by an elastothermic calculation under a given temperature produced by a thermal calculation, with an average temperature of around 200 °C. For the piston, the thermal factor produces significant distortion to the initially cylindrical form of the bearings. This is shown in Figure 3.20 for one of the piston bearings.

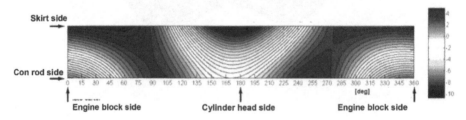

Figure 3.20. *Housing distortion in the piston bearings (μm)*

The lowest straight line of the cylinder (angle 0°, casing side) lifts up on the outer edge (skirt side) and lowers down on the inner edge (con rod side) while the highest straight line of the cylinder (cylinder head side, angle 180°) undergoes the opposite deformations. This modification of the form is introduced during the calculation as a flaw in the bearing's shape.

The diagrams of the load (load module) on the axis–piston and axis–connecting rod bearings for rotation frequencies of 13,500 and 18,500 rpm are shown in Figure 3.21. The increase in the engine speed causes a significant increase of the load on the bearings at crank shaft angle 0° (top dead center, start of the intake phase) due to the increase in inertial force exerted by the piston. At the second passing of the top dead center (end of the compression phase) this leads to an opposite effect, decreasing the load due to a high compensation of the compression forces. The reversals of the load at crank shaft angles 75° and 645° allow the bearings to be resupplied with lubricant.

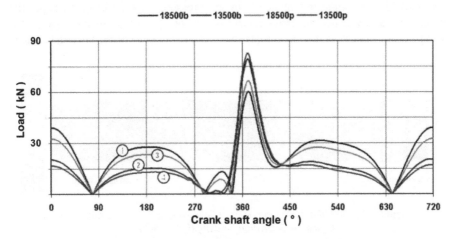

Figure 3.21. *Module of the load applied on axis-piston bearings (p) and on the con rod–axis bearing (b). Effect of piston inertia*

Among the "elementary solutions" determined to produce the compliance matrices is the one which corresponds to 1 MPa acting on the piston head surface. Figure 3.22 shows the deformation of the piston bearing's surface in one of the piston housings. The negative values obtained for bearing angles 0° and 180° are caused by a squeezing of the bearing of around 3 μm per MPa in the direction **x**, direction of the axis of the cylinder. In the orthogonal direction, an expansion of around 4 μm per MPa can be noted.

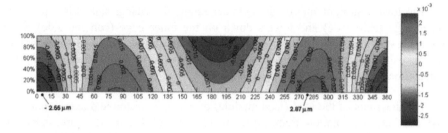

Figure 3.22. *Radial deformation (mm) of the piston–axis bearing housing for a 1 MPa pressure applied on the piston head surface*

Another "elementary solution" represents the effect of acceleration on the piston. Figure 3.23 shows the deformation of the surface of the bearings, precalculated by a standard field of 1 mm s^{-2}. It is also possible to observe an ovalization of the piston bearing which is similar to that caused by gas pressure when acceleration is positive (piston in the lower part of its trajectory) but becomes even more significant when acceleration is negative, the maximum values being obtained at the top dead center.

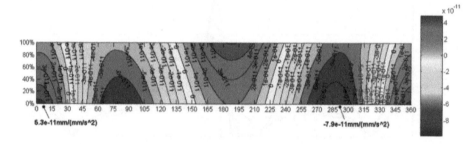

Figure 3.23. *Radial deformation (mm) of the piston–axis bearing housing for a 1 mms^{-2} acceleration field applied to the piston*

3.4.4.4. *Behavior at 20,000 rpm, without engine torque*

The first case study presented corresponds to a situation with maximum dynamic effect (maximum speed) and minimum effect due to gas pressure on the piston (negligible torque). Figure 3.24 shows the components of the load exerted on the connecting rod small end bearing. The absence of torque leads to a very low level of gas thrusts; the two parts of the engine cycle are, therefore, almost identical.

Under the effect of friction torques at the level of the two bearings, the axis acquires a rotation speed with respect to the piston which oscillates between 1,000 and 200 rpm as shown in Figure 3.25, producing a lack of symmetry which is not present in the movement of the connecting rod with respect to the piston.

This is due, in part, to the difference in friction coefficients between, on one side the axis and the connecting rod bush and, on the other side, the axis and the piston. Here, the former is greater than the latter. This behavior is more significant for the axis–connecting rod bearing due to the bending of the axis and the lower clearance level. Indeed, although the deformation of the axis is very beneficial for a better distribution of hydrodynamic pressure depending on the circumferential direction, bending deformation leads to metal-on-metal contact on the edges of the connecting rod bush. For the axis–piston bearing, however, this effect is not present since part of the operating clearance is greater due to heating and also the supporting part inside the piston can become slightly misaligned in order to take into account the deformation of the axis (on the connecting rod side, this rotation is almost impossible due to the proximity of the symmetry plane).

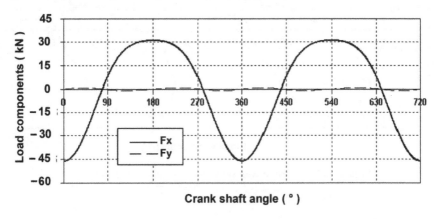

Figure 3.24. *Load applied on the connecting rod small end bearing at 20,000 rpm without engine torque*

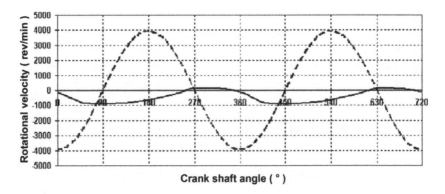

Figure 3.25. *Rotational velocities for the axis (———) and the connecting rod (– – –) with respect to the piston at 20,000 rpm*

The relation between the rotational speed of the axis and the total torque acting on it is shown in Figure 3.26. The torque is mainly due to friction in the connecting rod bearing (the torque in the piston bearings remains between –40 N mm and 25 N mm). At each rotation, two torque peaks can be observed of opposite values after the position of the maximum load at the top and bottom dead centers (shaft in compression or tension): the thickness of the film continues to reduce as long as the load remains significant, even if it drops off somewhat. Passing into mixed lubrication mode between the angles 0° and 40° then at 230° and 260° causes the dramatic increase in the friction torque.

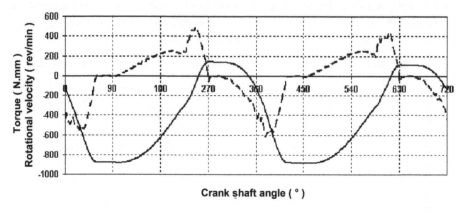

Figure 3.26. *Axis rotational velocity (——) and torque exerted on the axis (– – –) at 20,000 rpm*

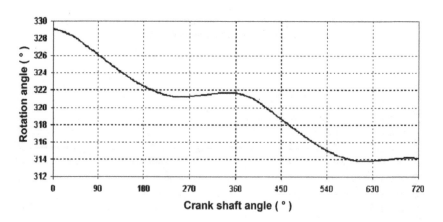

Figure 3.27. *Axis rotation angle at 20,000 rpm*

Figure 3.27 shows the evolution of the angular position of the axis in relation to a fixed reference point. At this operating speed, it can be observed that the rotation of the axis is in the direction opposite to that of the engine. The axis turns around 16° per engine cycle.

The minimum thickness of the film ((minimum (oil) film thickness) MOFT) is the parameter which best characterizes the behavior of a bearing in the case of EHD lubrication. Figure 3.28 shows the evolution of minimum thickness over an engine cycle for the two bearings of the piston–connecting rod link. The minimum thickness of the film is significantly lower for the small end bearing due to the bending of the axis and its ovalization as well as that of the small end connecting rod bush.

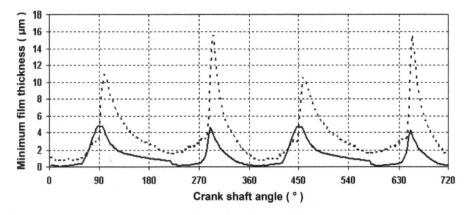

Figure 3.28. *Minimum film thickness at 20,000 rpm for axis–connecting rod bearing (———) and axis–piston bearings (– – –)*

The drawings of the pressure field in the link bearings at crank shaft angles 0° and 180° (Figures 3.29 and 3.30) show the areas of contact pressure located at the edges of the axis–connecting rod bearing, which is a result of the bending of the axis. At crank shaft angle 180°, the lubricant supply opens into the middle of the pressure field for the axis–connecting rod bearing, which emphasizes the contact pressure level at the edges of the bearing. These pressure fields confirm the absence of contact areas for axis–piston bearings in question, with the present running conditions.

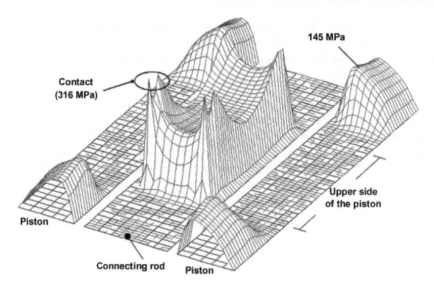

Figure 3.29. *Pressure fields at 20,000 rpm for the three bearings of the piston–connecting rod link at 0° crank shaft angle*

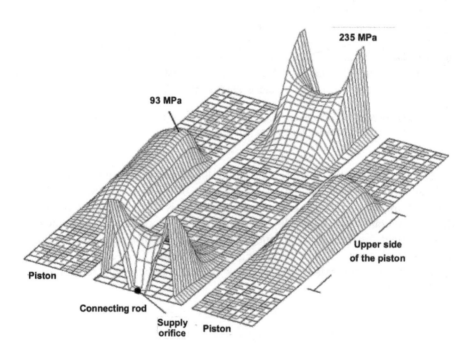

Figure 3.30. *Pressure fields at 20,000 rpm for the three bearings*

3.4.4.5. *Behavior at 17,000 rpm for the maximum engine torque*

For a Formula 1 engine, 17,000 rpm and maximum engine torque is representative of the harshest conditions that this type of engine can undergo. The diagram of the load exerted on the small end connecting rod bearing is shown in Figure 3.31 projected onto the basis of the connecting rod. As for the speed discussed in section 3.4.4.4 the transversal component of the load is negligible. In comparison with the previous case, it can be observed that the traction force exerted on the extremity of the connecting rod at crank shaft angle 0° is much lower and that at crank shaft angle 380°, the compression force is approaching 50 kN.

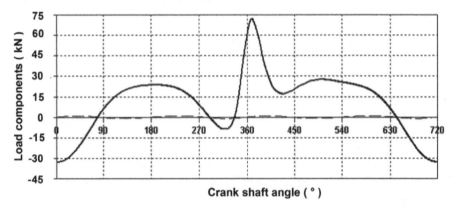

Figure 3.31. *Load on the connecting rod small end bearing at 17,000 rpm for the maximum torque*

Furthermore, it can be observed that, for an engine angle ranging from 90° to 630°, i.e. for three quarters of the cycle, the load is almost always on the same side, corresponding to a compression of the connecting rod. The angular speed of the axis in relation to the piston, shown in Figure 3.32, is above 1,000 rpm for three quarters of the cycle. It only reverses, and even then very slightly, for 90° of the crank shaft angle during the combustion phase. At the crank shaft angle 580° it reaches 2,400 rpm, which is a relatively high value when compared to the maximum angular speed of the connecting rod in relation to the piston (3,400 rpm). From this maximum point, the angular speed decreases slightly and then remains at the same value which is almost constant around 1,000 rpm until crank shaft angle 360°. Then, under the effect of the significant increase in load and the reduction in the thickness of the resulting film, the increase in friction due to the mixed lubrication conditions causes the drastic drop in speed, bringing it closer to the speed of the connecting rod.

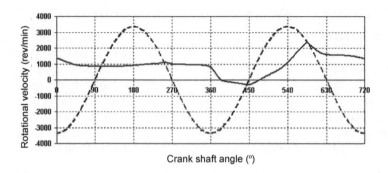

Figure 3.32. *Rotational velocity with respect to the piston for the axis (——) and the connecting rod (– – –) at 17,000 rpm*

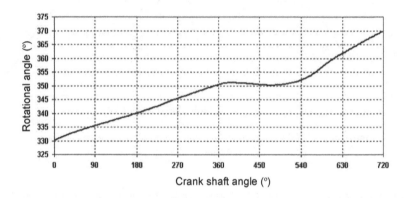

Figure 3.33. *Axis rotation angle at 17,000 rpm*

It can also be observed that the axis now turns in the same direction as the crank shaft (Figure 3.33), which is the opposite direction to the previous case, with only one "rest" period at around 380°–500°.

The evolutions along the cycle of the two friction torques exerted by the piston and the connecting rod on the axis are shown in Figure 3.34. The bending of the axis generates contacts on the edges of the connecting rod small end bush. These are always the main causes of the negative axis acceleration, even if in these operating conditions, the friction torque on the piston side momentarily reaches high values.

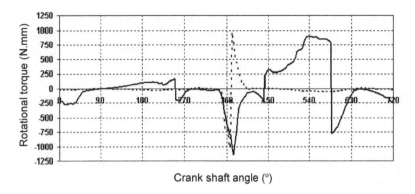

Figure 3.34. *Friction torques exerted on the axis by the connecting rod (———) and the piston (– – –) at 17,000 rpm*

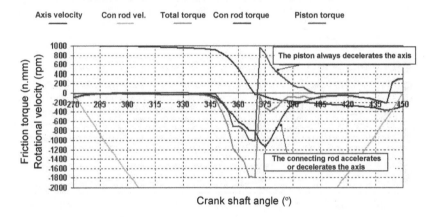

Figure 3.35. *Details of the friction torques exerted on the axis and the axis and connecting rod rotational velocities with respect to the piston*

Figure 3.35 shows the evolutions of the torques acting on the axis and the speeds of the axis and connecting rod with respect to the piston, thus enabling the analysis of the behavior of the axis in its "free" rotation. At each moment, its speed adjusts in order to equal the speed of the component which, at that time, is exerting the highest torque. Therefore, it tends to constantly reduce the dissipated energy. This method of saving dissipated energy would obviously be impossible for an arrangement where the axis is fixed either to the connecting rod or the piston and has a relative speed which is not "adjustable" but "imposed" by the other element.

The evolutions of the minimum film thickness for the two bearings are shown in Figure 3.36. It can be observed that, on average, the minimum thicknesses are greater for the axis–piston bearing. It can also be noted that the very short and weak double reversal of the load just before the start of the combustion creates a minimum thickness which is relatively high before a critical loading phase. Furthermore, these reversals of load favor the refeeding of the film for the axis–piston bearing.

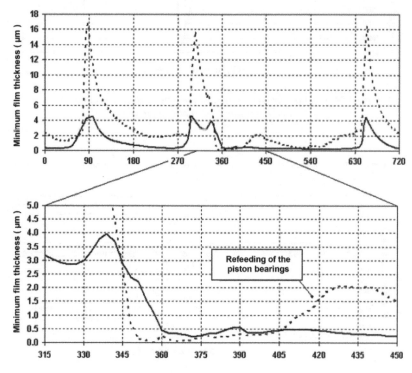

Figure 3.36. *Minimum film thickness at 17,000 rpm for connecting rod–axis bearing (——) and piston–axis bearings (– – –)*

The detail of the minimum film thickness in the area around the combustion top dead center can be seen at the bottom of Figure 3.36. The fact that the film of the axis–piston bearing is resupplied more easily is caused by a better feeding of lubricant at the bearing edges due to the higher operating clearance. The minimum thickness drops below 0.2 μm for both bearings.

The field of pressure (hydrodynamic and contact) for crank shaft angle 360° is shown in Figure 3.37. For the axis–piston bearings, the presence of contact pressure can be seen (hydrodynamic pressure is always zero on the edges of the bearing) due to the high load at this point. For the axis–connecting rod bearing, the bending of the axis always produces contact areas close to the edges of the bearing. As for the 20,000 rpm engine, the feeding opening in the middle of the bearing significantly reduces its load-bearing capacity.

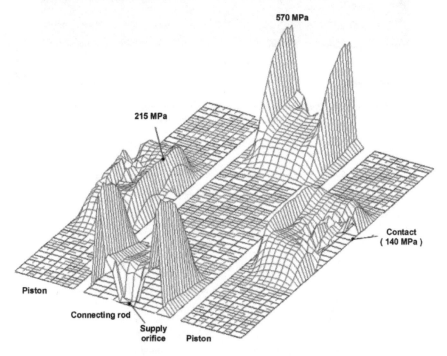

Figure 3.37. *Pressure fields at 17,000 rpm in the three bearings for the piston–connecting rod link at 360° crank shaft angle*

3.4.4.6. *Correction of the piston–axis bearing housing*

The numerical results given by the simulation at 17,000 rpm can be used to evaluate the wear level of the bearing machined in the piston. Figure 3.38 shows the distribution of the dissipated energy due to friction in the course of an engine cycle. As discussed previously, it is the edges of the bearings that are highly strained due to the bending of the axis during the combustion phase.

From the distribution field of dissipated energy, a correction of the surface profile of the piston bearings can be carried out. By establishing the goal of increasing the diametric clearance by at most 0.5 μm, it is possible to increase the local diameter by adding to the original profile, a profile obtained from dissipated energy data. The correction of the theoretical profile proposed is shown in Figure 3.39.

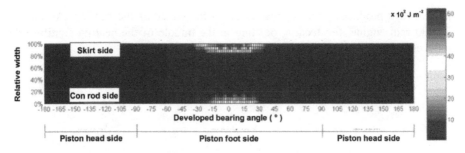

Figure 3.38. *Dissipated energy on the piston bearing surface during an engine cycle at 17,000 rpm*

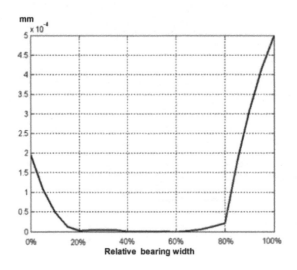

Figure 3.39. *Profile correction of piston bearing housings at 17,000 rpm*

3.4.4.7. *Influence of the deformation due to gas pressure*

The results of the calculations shown in section 3.4.4.3 have shown the importance of the deformation on the surface of the piston bearings due to the gas pressure acting on the piston head (Figure 3.22). We will now discuss the results of a simulation carried out by applying the same loads, but without taking into account the contribution of the deformation due to the gas pressure field.

The pressure fields in the three bearings of the link are shown in Figure 3.40. The crank shaft angle (369°) is slightly different from the angle corresponding to the fields in Figure 3.37 (360°) in order to show a localized contact area at the edge of the bearing on the connecting rod side (for crank shaft angle 360°, no contact is noted unlike when deformation due to gas pressure is taken into account, see Figure 3.37). The disappearance of the contact area on the outer edges of the piston bearings can also be observed.

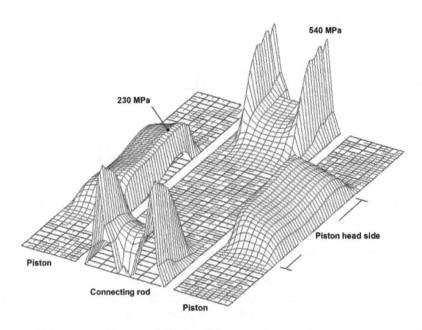

Figure 3.40. *Pressure fields at 17,000 rpm and 369° crank shaft angle, without considering the deformation due to gas pressure applied to the piston head*

This modification of the pressure fields is associated to a change in the minimum film thickness in the bearings, and mainly in the piston bearings. Figure 3.41 shows the comparison between the minimum thicknesses with and without the gas pressure effect. For the axis–connecting rod bearing, no change in the minimum thickness is observed. For the bearings machined in the piston, the deformation due to gas pressure has two harmful effects. First, this deformation prevents the second peak for lubricant refeeding toward crank shaft angle 340°. This peak, when it happens, follows the instantaneous canceling out of the load transmitted by the link. Its absence leads to a critical area of film thickness which starts sooner.

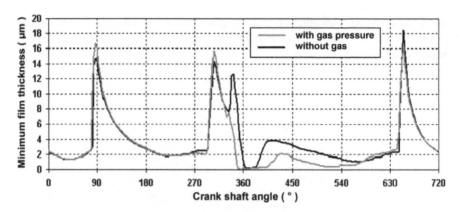

Figure 3.41. *Minimum film thickness at 17,000 rpm for the axis–piston bearings, with or without considering the deformation due to gas pressure*

The second harmful effect of the deformation caused by gas pressure on the piston head regards the refeeding after the combustion top dead center. Without such deformation, the refeeding starts almost immediately (around 20° after the top dead center), while in this case it is delayed by around a further 20°.

The result of these two combined events is that the period during which the thickness of the film is at a critical level, centered on the combustion top dead center, lasts around 50° of the crank shaft's rotation as opposed to only 20° in the other case. Since the harshest case takes the deformation caused by gas pressure into account, this highlights, once again, to what extent the definition of the model and the choice of parameters that it includes can influence the quality of the results obtained.

3.5. Bibliography

[BON 14a] BONNEAU D., FATU A., SOUCHET D., *Hydrodynamic Bearings* ISTE, London and John Wiley & Sons, New York, 2014.

[BON 14b] BONNEAU D., FATU A., SOUCHET D., *Mixed Lubrication Hydrodynamic Bearings*, ISTE, London and John Wiley & Sons, New York, 2014.

[OPT 00a] OPTASANU V., Modélisation expérimentale et numérique de la lubrification des paliers compliants sous chargement dynamique, Doctorate Thesis, University of Poitiers, France, 2000.

[OPT 00b] OPTASANU V., BONNEAU D., "Finite element mass-conserving cavitation algorithm in pure-squeeze motion, Validation/application to a connecting-rod small-end bearing", *Journal of Tribology*, vol. 122, pp. 162–169, 2000.

[OPT 99] OPTASANU V., BONNEAU D., "EHD lubrication of connecting-rod small end bearings using mass-conserving cavitation model", *Proceedings of 10th World Congress on the Theory of Machines and Mechanisms, Oulu University Press*, Finland, pp. 2496–2500, 1999.

[SPU 05] SPURIA M., BONNEAU D., LE BARATOUX Y., *et al.*, "A dynamic model for an internal combustion engine full floating piston pin in lubricated conditions", *Proceedings of 17th Congress of Italian Association of Theoretical and Applied Mechanics*, Florence, Italy, 2005.

[SPU 07] SPURIA M., Modellazione numerica delle condizioni di lubrificazione dello spinotto in un motore ad alte prestazioni, Doctorate Research Thesis in Car Design and Construction, University of Florence, Italy, 2007.

[ZIE 00] ZIENKIEWICZ O.C., TAYLOR R.L., *The Finite Element Method: The Basis*, 5th ed., Butterworth-Heinemann, Oxford, vol. 1, 2000.

4

The Engine Block–Crank Shaft Link

The link between the block and the crank shaft of an internal combustion engine, as well as the link between the block and the camshaft which is very closely related, are the most complex mobile arrangements since they involve a number of bearings whose lubrication parameters are interdependent. The complete models simultaneously consider the bearings as a whole with their relative displacements under the deformation effect of the engine block and the crank shaft. Even in the case of a crank shaft with a single crank pin (a single-cylinder or a V or a flat single-cylinder or a twin-cylinder engine), the two main bearings constitute, from the point of view of lubrication, a single bearing which must be considered in its entirety. Before addressing the description of the complete model, a greatly simplified model will be presented, which only takes into account one single journal bearing. This model, although in contradiction with what has been discussed above, is very similar to the one developed for the connecting rod big end bearing and can be used as a tool to preconfigure a main bearing. It will later be used to illustrate the specificities of the main bearings from the point of view of their lubricant supply.

Some of the developments discussed in this chapter are from the thesis of Thierry Garnier [GAR 97], and the later developments were carried out by Anne-Marie Chomat-Delalex [CHO 01, MOR 02].

4.1. Geometrical and mechanical particularities of the engine block – crank shaft link

Figure 4.1 shows a diagram of the link between an engine block and a crank shaft for an in-line four-cylinder engine. This link has five bearings whose ideal configuration is coaxial. Shown in the diagram are some of the mechanical actions exerted on the crank shaft:

– actions on the connecting rod at the level of the crank pins;
– centrifugal actions of the balance weights.

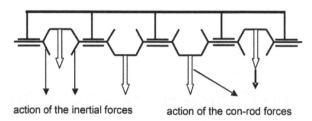

Figure 4.1. *Schematic of the engine block – crank shaft link*

The balance of the crank shaft is ensured by the linking forces, which operate on its five bearings. In the developments discussed below, axial forces are not taken into account and are not shown in the diagram, the balance of which is ensured by a thrust bearing. Since the forces exerted by the connecting rods are not synchronized, the reactions at the level of the five bearings differ, and the crank shaft becomes misaligned. If the crank shaft and engine block are considered as being totally undeformable, the angle of misalignment is limited by the clearance available in the bearings and in particular, in the two bearings located at the extremities. In reality, under the intensity of the forces undergone, the crank shaft and, less so, the engine block experience deformation. The misalignment is, therefore, significantly greater and different for each bearing.

For each bearing, the average position of the journal bearing in relation to the average position of the bearing is given by the eccentricity parameters ε_x and ε_y and misalignment ζ_x and ζ_y (Figure 4.2). These reference positions are defined as "average" because, under the effect of hydrodynamic and possibly contact pressures, the cylindrical surfaces of the bearing shaft and sleeves become deformed. At the same time, the entire structure also deforms, which requires extra parameters for the positioning of the journal bearing and the bearings in relation to each other. This last point is detailed later during the description of complete models.

4.2. Lubricant supply

The supply of a main journal is carried out by a duct drilled into the engine block (Figure 4.3). This duct opens out into a semi-cylindrical groove. This is used to supply the crank pin bearings with a duct drilled diametrically in the journal bearing, so that one of the duct's apertures always opens out onto the groove. Inclination φ of the duct openings in relation to the journal bearing when the engine is at top dead center must be chosen with consideration so that the hydrodynamic pressure field is unaffected as much as possible by the presence of these openings.

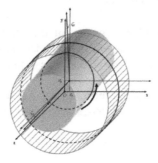

Figure 4.2. *Position parameters of the crank shaft relative to the engine block*

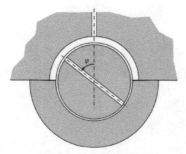

Figure 4.3. *Lubricant supply of a main bearing*

The finite element modeling of the thermoelastohydrodynamic (TEHD) problem for the journal bearing uses a film mesh of quadrangular elements. Although it is easy to make the mesh coincide with the edges of the semi-cylindrical groove, the same cannot be said for the circular openings. For these, the operation is limited to imposing a supply pressure only to the nodes, which happen to coincide with the surface they are occupying at each moment.

4.3. Calculus of an isolated crank shaft bearing

For this example, we will consider the case of a main bearing of an average-sized diesel engine from a standard car. The calculation is carried out in isotherms, and it is assumed that the crank shaft is rigid.

The bearing and its finite element mesh are shown in Figure 4.4. Since the housing cap bears the strain from compression forces, its form is defined in order to give it a higher mechanical resistance.

For calculating the compliance matrix of the journal bearing, it is held by a clamp placed at the level of the base of the cylinder liner, at a large enough distance from the bearing for it not to have an effect on its deformation.

The main characteristics of the bearing, as well as the lubricant, are given in Table 4.1. The lubricant has piezoviscous behavior which is governed by the Barus law:

$$\mu = \mu_0 e^{\alpha(p-p_0)}$$

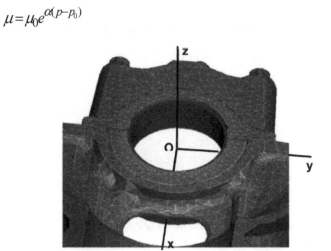

Figure 4.4. *References and mesh of a main bearing*

Main bearing diameter	49	mm
Bearing width	19	mm
Radial clearance	20	μm
Rotational frequency	2000	tr/min
Viscosity μ_0 of the lubricant at ambient pressure	0.004	Pa.s
Piezoviscosity coefficient α	0.011	MPa^{-1}
Supply pressure	0.4	MPa

Table 4.1. *Main parameters of the bearing and of the lubricant*

Figure 4.5 shows the components of this bearing's load for a speed of 2,000 rpm and for a crank shaft, which is statically balanced[1]. The force is essentially shown by component F_x, with a negative sign, oriented toward the housing cap. It can be observed that the two thrusts at crank shaft angles 370° and 550° correspond to the combustion phases in the cylinders located on both sides of the bearing in question.

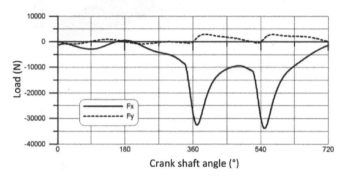

Figure 4.5. *Load at 2,000 rpm. Crank shaft bearing*

The representation of the pressure field at crank shaft angle 90° (Figure 4.6) shows the role played by the supply groove and the two diametrically drilled openings. In the part where an increase in pressure develops due to the hydrodynamic effect, the pressure field is reduced by the groove. Due to the presence of the groove, there is an excellent lubricant supply to the bearing. Figure 4.7 shows the leakage flow at one of the bearing edges at crank shaft angle 360° for the two configurations studied (diametric drilling at 90° and 135°). At this moment, the thrust in the direction of the housing cap is strong, and the area of film with low thickness is on this side (between angles 90 and 270° of the developed bearing). The leakage flow at this point is, therefore, weak. However, on the other side of the bearing (between angles 0 and 90° and then between angles 270 and 360°) the thickness of the film is higher, and a significant amount of lubricant passes from the groove to the edge of the bearing. At these angles, the orientation of the drilling has little effect on the distribution of the leakage.

---

1 For a crank shaft, which is not statically balanced, it is necessary to add the projections of the centrifugal force caused by imbalance to the components F_x and F_y. This point is described in detail by the complete model (see section 4.4).

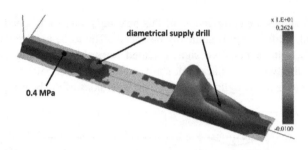

Figure 4.6. *Hydrodynamic pressure field at 200° crank angle. Supply drill at 90°*

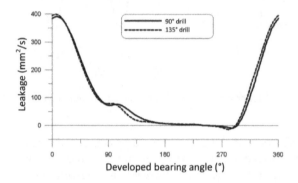

Figure 4.7. *Oil leakage on the bearing edge at 360° crank shaft angle*

Due to favorable supply conditions, the flow of the lubricant leaving the bearing is relatively high. Figure 4.8 shows its evolution during the engine cycle for the configurations studied (diametric drilling at 90° and 135°). It can be observed that the re-entering flow at the extremities of the bearing is almost zero, only present for a short moment around angle 200°. Indeed, there is a slight input on the distribution of the leakage of Figure 4.7, located between angles 230 and 300° of the developed bearing.

The flow averaged over a cycle amounts to 0.62 dm^3/min. For the connecting rod big end bearing of the diesel engine studied in section 2.6, whose dimensions are similar (44 mm in diameter and 18 mm wide) and rotating at the same speed, but supplied by a single opening (supply pressure 0.3 MPa), a flow of 0.1 dm^3/min is obtained.

The importance of the main bearing flow helps it to obtain better cooling. The thermohydrodynamic calculation of this bearing, therefore, appears to be less necessary than it is for a connecting rod big end bearing.

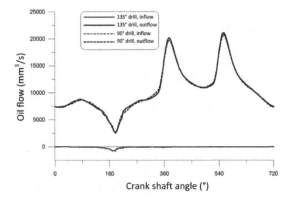

Figure 4.8. *Oil inflow and outflow on the bearing edge*

Due to the high level of load being supported, the bearing with a sleeve reference profile, which is assumed to be perfectly cylindrical, works in conditions of mixed lubrication. Figure 4.9 shows the evolution of the maximum contact pressure during the engine cycle for a sleeve with its actual original profile (marked "new drill") and for a sleeve with a worn profile. This profile is obtained after 10 cycles to calculate the form adjustment (wear model 1 as described in section 4.4.1 of [BON 14a]). For the sleeve with the worn profile, the maximum contact pressure is significantly reduced. Furthermore, it can be observed that the evolutions of the maximum contact pressure are very different for the two drilling angles of the diametric duct. For the "new" sleeve, the three maximums are clearly more spaced out when the drilling is oriented at 90°, the central maximum is higher and the two others are lower than for the orientation at 135°. This difference will lead to different levels of wear for the sleeves.

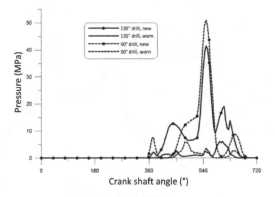

Figure 4.9. *Contact pressure*

An observation of the contact pressure fields (Figure 4.10) obtained for the sleeve with a "new" profile allows us to understand the cause of this difference. The fields are relative to crank shaft angle 610°, an angle where the maximum contact pressures vary the most (Figure 4.9). For the drilling at 90°, the maximum contact pressure is at the edges of the bearing. However, at 135° it is at the edges of the supply opening, which at this moment is in the area of minimum thickness and maximum hydrodynamic pressure. The weakening of the bearing housing in the low film thickness zone due to this duct is compensated by a rise in contact pressure. The –45° displacement of the opening (drilling at 90°) helps to avoid this problem. The positioning of the diametric drilling remains a delicate issue: displacement can improve elastohydrodynamic (EHD) behavior at certain moments in the cycle (as in this example) but worsen the situation at other moments, especially with a simultaneous displacement of the second opening, which can in turn enter into a critical zone.

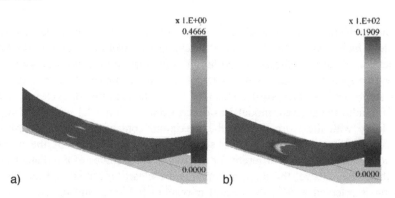

Figure 4.10. *Film thickness and contact pressure at 610° crank angle: a) diametrical drill at 90° and b) diametrical drill at 135°*

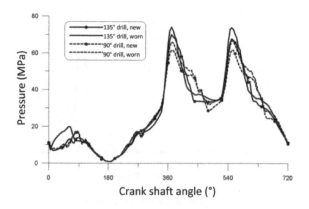

Figure 4.11. *Hydrodynamic pressure*

The evolution of the maximum hydrodynamic pressure is less sensitive to the placement of the diametric drilling (Figure 4.11). The two curves which correspond to the drilling at 90° are very similar. The same can be said for the two curves representing the drilling at 135° except in the area located between the angles 0 and 80°, where the maximum hydrodynamic pressure for the worn sleeve is clearly higher to those obtained in the three other cases. Since there is no contact pressure in this phase (Figure 4.9), the only difference that can arise comes from the difference between the worn forms obtained for the two drilling configurations.

The evolutions of minimum film thickness are significantly less affected by the change in the duct's orientation (Figure 4.12).

Figure 4.13 shows the fields of wear after 10 calculation cycles for the two drilling configurations of the supply duct. A comparative analysis of these two fields shows that for the drilling at 90°, the wear is more significant on the edges of the bearing of around 0.3 µm. Nevertheless, for the drilling at 135°, the wear is slightly more apparent in the central area of the bearing due to the passing of the duct openings and the contact areas which surround them (Figure 4.10).

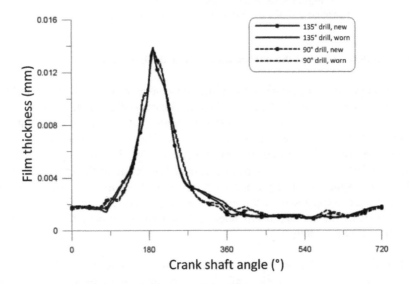

Figure 4.12. *Minimum film thickness*

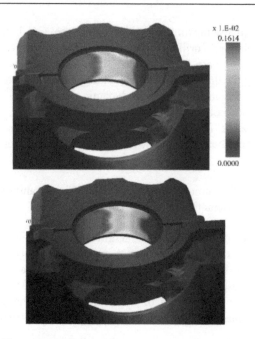

Figure 4.13. *Bearing sleeve wear: (top) drill at 90°. Maximum 1.614 μm; (bottom) drill at 137°. Maximum 1.319 μm*

4.4. Complete model of the engine block – crank shaft link

Excluding the single-cylinder engine, the engine block – crank shaft link is statically indeterminate: the forces on the main bearings depend on the loads exerted by the connecting rods at the level of the crank pins but also on the deformations of the crank shaft and, less so, of the engine block. This interdependence requires the development of models which simultaneously consider all of these parameters. The model detailed below was developed by Thierry Garnier in his thesis work [GAR 97, GAR 99]. Examples of application of this model along with validation tests, especially for comparisons with experimental results, can be found in various publications by the authors of this book and their collaborators [BON 99, BON 00, CHO 01, MOR 01, MOR 02].

For the single-cylinder engine, symmetry considerations lead to the assumption that the engine block – crank shaft link is isostatic. It is, therefore, possible to consider the two journal bearings, located on both sides of the single crank pin, as a single bearing with a very large complete central groove occupying all the space including the area between the main bearings and where there is an ambient pressure. The model of the isolated main bearing (section 4.3) can, therefore, be directly applied. This model can also be used for the calculation of the (T)EHD behavior of the spindle bearings [FAT 05a, FAT 05b].

4.4.1. *Model presentation*

For a link with $nb + 1$ bearings, the engine block and the crank shaft are subdivided into nb "segments", each one ranging from bearing i to bearing $i + 1$. Figure 4.14 shows segment i, in a projected view on the plane O z x. The reference frame center is point O, the center of the housing of the first bearing without deformations, and the reference frame axis O z is an axis starting from O and passing through the center of the housing of the final bearing, without deformations. The housing centers of the intermediate bearing are not necessarily placed on O z (in the case of flaws in the initial forms). Each housing has an axis of revolution. In the event of misalignment, these axes are not parallel to the axis O z. All these flaws are quantified by 4 $(nb + 1)$ parameters: two positioning parameters and two angular parameters per bearing.

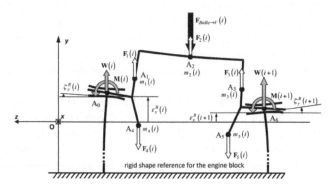

Figure 4.14. *Block and crank shaft distortion and mechanical actions*

Under the effect of the forces applied and the thermal deformations, the bore centers move. Their positions are given by the parameters $\varepsilon_x^B(i)$ and $\varepsilon_y^B(i)$, $i = 1$, $nb + 1$. In the same way, the orientation of the bearings changes and is given by $\zeta_x^B(i)$ and $\zeta_y^B(i)$, $i = 1$, $nb + 1$. The parameters $\varepsilon_y^B(i)$ and $\zeta_x^B(i)$ corresponding to the off-plane displacements are not shown in Figure 4.14.

Each segment i of the crank shaft is modeled by a seven-node element as shown in Figure 4.14. These nodes are connected up by flexible subelements. The nodes are assigned a mass and an inertia matrix. The masses $m_j(i)$ of the five nodes, which are not located on the rotation axis, are used in the calculation of the centrifugal forces undergone by the crank shaft. The center point for the reference frame of the crank shaft is O_1, the center of the first crank pin (bearing 1 of segment 1) and the axis $O_1 z_1$ is the straight line linking O_1 to the center of the last crank pin without any deformation. As for the housings on the engine block side, each crank pin has four initial positioning parameters which may be zero if there is no initial misalignment. Each node j of segment i has four more displacement parameters $\varepsilon_x^V(j,i)$, $\varepsilon_y^V(j,i)$, $\zeta_x^V(j,i)$ and $\zeta_y^V(j,i)$ resulting from the thermoelastic deformation of the crank shaft. The translation and rotation displacements relative to direction z are of little interest except for the node located at the center of bearing nb. In order to take into account the torsion deformation of the crank shaft under the effect of the torque exerted by the timing belt and/or accessories, this node is assigned a rotation parameter ζ_z around axis $O_1 z$.

The forces which act on the crank shaft segment (apart from internal forces exerted by neighboring segments) are also shown in Figure 4.14. Forces $\mathbf{F}_{\text{Con rod}\rightarrow\text{Pin}}(i)$ exerted by the connecting rod (or the two connecting rods in the case of a V-shape engine) as well as centrifugal forces $\mathbf{F}_j(i), j = 1, 5$, are known beforehand. The resultants $\mathbf{W}(i)$ and moments $\mathbf{M}(i)$ exerted by the pressure fields are not known. All of these forces and consequential displacements must satisfy the equilibrium equations of the structure and elastic behavior of each of the elements in question.

Assuming that $\mathbf{T}_E(i)$ is the torsor of external actions exerted on segment i of the crank shaft, the resultant of which is the sum of the driving force $\mathbf{F}_{\text{Con rod}\rightarrow\text{Pin}}(i)$ transmitted by the connecting rod to the crank pin and the centrifugal forces $\mathbf{F}_j(i)$, $j = 1, 5$. The resulting moment of this torsor is the sum of the moments caused by each of these forces. This moment is expressed at crank shaft reference point O_1.

Assuming that $\mathbf{T}_p(i)$ is the torsor of the pressure actions exerted on the main bearing i of the crank shaft, with the resultant $\mathbf{W}(i)$ and resultant moment $\mathbf{M}(i)$ expressed in O_i, center of the journal bearing i. The components of these resultants and moments are calculated with the relations:

$$\Im_{pression} \begin{cases} W_x(i) = R_i \iint_{\Omega(i)} p_i(\theta,z)\cos\theta\, d\theta dz \\ W_y(i) = R_i \iint_{\Omega(i)} p_i(\theta,z)\sin\theta\, d\theta dz \\ M_x^{O_i}(i) = -R \iint_{\Omega(i)} z\, p(\theta,z)\sin\theta\, d\theta dz \\ M_y^{O_i}(i) = R \iint_{\Omega(i)} z\, p(\theta,z)\cos\theta\, d\theta dz \end{cases}$$

[4.1]

where R_i is the radius of the bearing i and θ is the angle of the developed bearing.

The EHD behavior of the engine block – crank shaft link depends on the elastic behavior of the crank shaft and the engine block. This behavior must be taken into account at the time of carrying out the calculation. On the one hand, the elastic behavior of the crank shaft is described in a global sense by the four displacement parameters of each of its 4 (6 nb +1) condensation nodes, while on the other hand, in a local sense by the radial displacement of each of the surface nodes for each main bearing. In the same way, the elastic behavior of the engine block is described in a global sense by the four parameters $\varepsilon_x^B(i)$, $\varepsilon_y^B(i)$, $\zeta_x^B(i)$ and $\zeta_y^B(i)$ for the positioning of each bearing and locally by the radial displacement of each of the nodes on the surface of each sleeve.

4.4.2. Expression of the elastic deformations

After the assembly of the nb segments, a stiffness matrix and a mass matrix of 4 (6 nb +1) + 1 can be obtained for the whole of the crank shaft arrangement. These matrices can be obtained by the condensation of global matrices calculated with a three-dimensional (3D) finite element model of the crank shaft or from a frame model of the crank shaft using the method developed and validated by Hodgetts [HOD 74]. The first node of the crank shaft located at the center of the first main bearing is chosen as the reference node. This enables us to reduce the stiffness matrix order to 4 and make it reversible. The reversal of the matrix obtained in this way gives the global compliance matrix of the crank shaft.

The equation used to translate the global elastic behavior of the crank shaft, in the generalized form, is written as follows:

$$\mathbf{q}_S = [S_S]\, \mathbf{T}_S \qquad [4.2]$$

where $\mathbf{q}_S$ represents the generalized vector of global displacement of the crank shaft, relative to the first node located at the center of the first crank pin, made up of the 4 (6 nb + 1) + 3 displacement components of the crank shaft's condensation nodes, $[S_V]$ is the global compliance matrix of the crank shaft and $\mathbf{T}_S$ is the vector of the loads (resultants and moments) applied. $\mathbf{T}_S$ is obtained by the assembly of nb vectors $\mathbf{T}_E(i)$ and $\mathbf{T}_p(i)$ as well as the torsion torque of the crank shaft, followed by a reduction, since the first node of the crank shaft is used as the reference node.

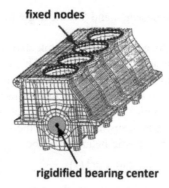

Figure 4.15. *Finite element model of the engine block*

The overall displacement of the bearings is also expressed by a compliance matrix of order 4 ($nb + 1$). In order to determine this matrix, a finite element model of the engine block is used, in which each housing is "filled" by a "rigidified bearing center" (Figure 4.15). After having imposed the clamping conditions on the model (in practice, the cylinder head plane is chosen as the rigid reference surface and is assigned clamping conditions), four separate forces – a force following **x**, a force following **y**, an **x** axis torque and finally, a **y** axis torque – are applied successively to the center of each "rigidified bearing center". For each of the loads, after deformation calculations, displacements following **x** and **y** are recorded as well as the rotations relative to the same axes for the $nb +1$ "rigidified bearing centers". In this way, the overall compliance matrix of the engine block can be obtained.

The equation which translates the overall elastic behavior of the engine block, in a generalized form, is written as follows:

$$\mathbf{q}_B = [S_B] \, \mathbf{T}_B \qquad [4.3]$$

where $\mathbf{q}_B$ represents the generalized displacement vector made up of the 4 ($nb + 1$) displacement components for the center of the $nb + 1$ bearings, $[S_B]$ is the compliance matrix of the engine block and $\mathbf{T}_B$ is the vector of loads (resultants and moments) applied to the center of each bearing. $\mathbf{T}_B$ is obtained by the assembly of $nb + 1$ vectors $-\mathbf{T}_p(i)$ (the minus sign is introduced, since the relations [4.1] express the forces exerted by the pressure field on the sleeve).

The local radial displacements are calculated for each bearing using the compliance matrices obtained following an identical process to the isolated bearing described in section 4.3. When the journal bearings and/or the structures surrounding the bearings are identical, a single set of compliance matrices can be used for all of the bearings.

4.4.3. Expression of the film thickness

For each bearing of the link, the film thickness at a node j of the surface mesh of the bearing is written as follows:

$$h_j = h_N\left(\theta_j, z_j\right) + h_D\left(\theta_j, z_j\right) + h_E\left(\theta_j, z_j\right) + h_T\left(\theta_j, z_j\right) \qquad [4.4]$$

where h_N represents the nominal thickness, h_D is the extra thickness resulting from a flaw in the form, h_E is the extra thickness due to the elastic deformation of the surfaces of the crank pin and the housing and h_T is the extra thickness due to thermal deformations of these surfaces.

The nominal thickness involves not only the radial clearance C but also the position of the center of the shaft in the bearing as well as its misalignment as shown in Figure 4.2. By applying the small-angle approximation, this thickness can be written as follows:

$$h_N(\theta, z) = C\left(1 - \varepsilon_x \cos\theta - \varepsilon_y \sin\theta\right) - \left(\zeta_y \cos\theta - \zeta_x \sin\theta\right) z \qquad [4.5]$$

For bearing i, the vector of elastic components $\mathbf{h}_E(i)$ is given by (see Chapter 4 of [BON 14b]):

$$\mathbf{h}_E(i) = [C(i)]\, \mathbf{p} \qquad [4.6]$$

where $[C(i)]$ is the combined compliance matrix of the surfaces of the crank pin and the housing (for the housing, a rotation operation may be necessary, see section 4.3 of [BON14b]) and $\mathbf{p}$ is the vector of nodal pressures.

The calculation for thermal deformation is described in [BON 14c].

4.4.4. Equation system

The equation system that represents the EHD[2] behavior of the engine block – crank shaft link is made up of:

– the Reynolds equation governing the hydrodynamic behavior of the lubricant, in each of the bearings;

---

[2] The model described is in isotherms. In the event of TEHD processing, it is advisable to add the equations of one of the thermal models described in [BON 14c].

– the equations of mixed lubrication;

– equilibrium equations for the crank shaft;

– equations governing the elastic behavior of the crank shaft;

– equations governing the elastic behavior of the engine block;

– equations due to the hyperstaticity of the mechanism and translating the compatibility of displacements from the center of one bearing to the center of the next.

The Reynolds equation and its numerical processing are described in detail in Chapters 2 and 3 of [BON 14b], and the mixed lubrication equations are detailed in [BON 14a]. The equations governing the elastic behavior of the engine block and the crank shaft and their numerical processing are described in Chapter 4 of [BON 14b] and in the previous section.

Ignoring vibratory phenomena and considering centrifugal forces as actions external to the crank shaft, the static equilibrium of the crank shaft can be written in the following generalized form:

$$\mathbf{T}_E + \mathbf{T}_p = 0 \qquad [4.7]$$

with:

$$\mathbf{T}_E = \begin{cases} \sum_{i=1}^{np-1}\left(F_{B\to V}(i) + \sum_{j=1}^{5}F_j(i)\right) + F_{A\to V} \\ \sum_{i=1}^{np-1}\left(O_1A_2(i)\wedge F_{B\to V}(i) + \sum_{j=1}^{5}O_1A_j(i)\wedge F_j(i)\right) + M_{O_1}(F_{A\to V}) + C \end{cases} \qquad [4.8]$$

$$\mathbf{T}_p = \begin{cases} \sum_{i=1}^{np} W(i) \\ \sum_{i=2}^{np}\left(O_1A_0(i)\wedge W(i) + M(i)\right) + \sum_{i=1}^{np} M(i) \end{cases} \qquad [4.9]$$

$\mathbf{F}_{C\to S(i)}$ represents the forces $\mathbf{F}_{Con\,rod\to Pin}(i)$ exerted by the connecting rod of the crank shaft and $\mathbf{F}_{A\to V}$ is the resultant of the other forces exerted on the crank shaft, such as, for example, those coming from the timing belt and/or the accessory belt. $\mathbf{C}$ represents the torque reacting to the engine torque exerted at the end of the crank shaft, on the gearbox side. The friction torques in each of the bearings are negligible compared to other forces and are not taken into consideration. For example, for a medium-sized engine, the torque exerted by a connecting rod can reach 2.5 kN.m while the friction torque does not exceed 0.1 N.m.

To take the vibratory behavior of the crank shaft into consideration, it is necessary to define generalized external forces $\mathbf{T}_E$, the forces and moments due to the linear and angular accelerations of condensation mass.

The displacement parameters of the crank shaft nodes and bearing centers are defined separately. Nevertheless, it is necessary to ensure the compatibility of these displacements: for each loop of the linkage from bearing i to bearing $i + 1$ (Figure 4.16), the algebraic sum of the positioning parameters must be zero. For each bearing i, the crank pin is localized by parameters $\varepsilon_x(i)$, $\varepsilon_y(i)$, $\zeta_x(i)$ and $\zeta_y(i)$. If $\varepsilon(i)$ is noted as the vector of these four parameters, the compatibility relationship between these displacements can be written as follows:

$$\mathbf{q}_B(i+1) - \mathbf{q}_B(i) + \varepsilon(i+1) - \varepsilon(i) + \mathbf{q}_{V6}(i) - \mathbf{q}_{V0}(i) + \mathbf{dr}(i) = 0 \qquad [4.10]$$

where $\mathbf{dr}(i)$ represents the displacement of the main bearing $i+1$ due to the rotation of the crank pin i in bearing i whereby:

$$\mathrm{dr}(i) = \left(z_6(i) - z_0(i)\right)\begin{Bmatrix} -\zeta_y(i) & \zeta_x(i) & 0 & 0 \end{Bmatrix}^T$$

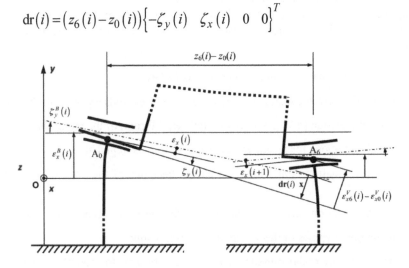

Figure 4.16. *Displacement compatibility*

4.4.5. Resolution method

Like for a single bearing model, the equations of the system are nonlinear. They are solved through a repetitive Newton–Raphson-type process. The complete system of equations can be written as:

$$\mathbf{R} + [J]\,\Delta = 0 \qquad [4.11]$$

where **R** is the vector of the residuals of the system equations, [J] is the Jacobian matrix and Δ is the unknown vector to be determined at each time step.

The calculation of the Jacobian matrix is at the heart of the Newton–Raphson process, since this matrix contains the partial derivatives of all the equations of the system in relation to all the unknown parameters. In this case, due to the number of bearings in the link, nb, the order of this matrix is large:

$$\sum_{i=1}^{nb} N_i + 12\,nb - 4$$

where Ni is the number of nodes on the surface mesh of the bearing i.

However, the structure of the Jacobian matrix has a number of zero-value components. This structure can be drawn upon to optimize the time taken for the calculation. The linear system is resolved in blocks, via submatrices as described below.

The equations of the system are denoted as follows:

– E_i: all discretized equations of the generalized Reynolds equation for bearing i;

– E_{SE}: all equation [4.7] given by the static equilibrium of the crank shaft;

– E_C: all equations for displacement compatibility as in [4.10];

– E_{EB}: all equations translating the elastic behavior of the engine block, as in [4.3];

– E_{ES}: all equations translating the elastic behavior of the crank shaft, as in [4.2].

The unknown values of the system are the following:

– $\mathbf{p}_i$: vectors of nodal pressures (or lubricant fillings for complementarity) for bearing i;

- ε_i: vector of the positions of the crank pin in each bearing;
- $\mathbf{q}_B$: vector of the positions of the bearings, on the engine block side
- $\mathbf{q}_S$: vector of positions relative to pin 1 of the other pins of the crank shaft.

Unknowns equations	$\mathbf{p}_1$	...	$\mathbf{p}_{nb}$	ε_1	...	ε_{nb}	$\mathbf{q}_B$	$\mathbf{q}_S$
E_1	X			X				
...		X			X			
E_{nb}			X			X		
E_{SE}	X	X	X					
E_C				X		X	X	X
E_{EB}	X	X	X				X	
E_{ES}	X	X	X					X

Table 4.2. *Equations and variables identification*

Table 4.2 identifies the variables that appear in each equation.

The Jacobian matrix can, therefore, be broken down in the following way:

$$[J] = \begin{bmatrix} [A_{11}] & [0] & \cdots & [0] & [A_{1m}] \\ [0] & [A_{22}] & \cdots & [0] & [A_{2m}] \\ \vdots & \vdots & \ddots & \vdots & \vdots \\ [0] & [0] & \cdots & [A_{nb\,nb}] & [A_{nb\,m}] \\ [A_{m1}] & [A_{m1}] & \cdots & [A_{m\,nb}] & [A_{mm}] \end{bmatrix} \quad [4.12]$$

where each submatrix is defined as follows:

$$[A_{k,k}] = \begin{bmatrix} \partial E_k / \partial p_k \end{bmatrix}$$

$$[A_{m,k}] = \left[\left[\frac{\partial E_{SE}}{\partial p_k}\right] [0] \left[\frac{\partial E_{EB}}{\partial p_k}\right] \left[\frac{\partial E_{ES}}{\partial p_k}\right]\right]^T$$

$$[A_{k,m}] = \left[[0] \quad \cdots \quad \left[\frac{\partial E_k}{\partial \varepsilon_k}\right] \quad \cdots \quad [0] \quad [0] \quad [0]\right]$$

$$[A_{m,m}] = \begin{bmatrix} [0] & \cdots & [0] & \cdots & [0] & [0] & [0] \\ \left[\frac{\partial E_C}{\partial \varepsilon_1}\right] & \cdots & \left[\frac{\partial E_C}{\partial \varepsilon_k}\right] & \cdots & \left[\frac{\partial E_C}{\partial \varepsilon_{nb}}\right] & \left[\frac{\partial E_C}{\partial q_{EB}}\right] & \left[\frac{\partial E_C}{\partial q_{ES}}\right] \\ [0] & \cdots & [0] & \cdots & [0] & \left[\frac{\partial E_C}{\partial q_{EB}}\right] & [0] \\ [0] & \cdots & [0] & \cdots & [0] & [0] & \left[\frac{\partial E_C}{\partial q_{ES}}\right] \end{bmatrix}$$

4.4.6. *Examples*

In order to provide examples, we present several simulations related to an in-line four-cylinder engine. The numbering convention for the bearings is given in Figure 4.17. The rotation of the crank shaft follows Z, in the trigonometric sense.

Table 4.3 gives the main parameters of the main bearings and the lubricant. The lubricant has piezoviscous properties, which are governed by the Barus law.

Figure 4.18 shows the components of the load applied to the crank pins for speeds of 2,500, 4,000 and 6,000 rpm. The order of ignition is 1, 3, 4 and 2, and the ignition angles are, respectively, 0, 180, 360 and 540°. It can be observed that at 6,000 rpm, the forces due to the inertia of the connecting rod and the piston fully compensate the forces caused by the gas thrust.

The calculations at 2,500 rpm are carried out with ("elastic" case) and without ("rigid" case) consideration for the local elasticity at each main bearing. In the rigid case, the maximum eccentricity is logically below 1 (see Figure 4.19). The elasticity of the bearings gives extra flexibility to the system, which leads to more significant displacement of the shaft within the bearings.

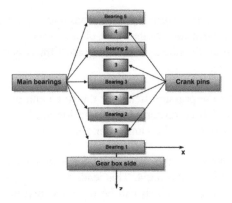

Figure 4.17. *Bearings numbering convention*

Bearing diameter	49	mm
Bearing width	19	mm
Radial clearance	30	µm
Rotational frequency	2,500; 4,000; 6,000	rpm
Lubricant viscosity	0.007	Pa.s
Piezo-viscosity coefficient α	0.011	MPa^{-1}
Supply pressure	0.2	MPa

Table 4.3. *Main bearing and lubricant data*

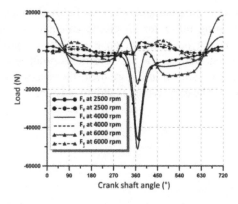

Figure 4.18. *Load at 2,500, 4,000 and 6,000 rpm. Con-rod big end bearings*

As can be observed in Figure 4.20, the maximum pressure predicted with a "rigid" calculation is much greater than that obtained with an "elastic" calculation. However, the maximum thickness of the lubricant film is generally greater for the rigid case (Figure 4.21).

Figure 4.22 shows the pressure fields for the five main bearings, obtained at the crank shaft angles, which correspond to the ignition angles. Figure 4.23 shows the fields of thickness obtained for the same crank shaft angles.

At crank shaft angle 0°, the maximum force applied to the crank shaft is located at the level of crank pin no. 4. This leads to significant pressure at the level of main bearings no. 4 and no. 5. It can be observed that the pressure field is not symmetrical, which indicates a misalignment of the crank shaft in the housings. At crank shaft angle 180°, the maximum force is located at the level of crank pin no. 2. It is interesting to observe that maximum pressure is obtained at the level of bearing no. 1. Generally speaking, the two bearings, which suffer the most severe conditions, are no. 1 and no. 5 located at the two extremities of the crank shaft.

The analysis of the thickness fields shown in Figure 4.23, shows an overall swing of the crank shaft between the angles of 0 and 360°, generated by a significant force at the level of crank pin no. 1 at 0° and no. 4 at 360°. At crank shaft angles 180 and 540°, the maximum force is exerted on crank pins no. 2 and no. 3, which take up a more central position in the crank shaft. Thus, the crank shaft has a displacement in the direction of the load, which is similar to that of a rigid element, which leads to equal minimum thicknesses for all of the main bearings. In Figure 4.21, which compares the maximum pressures obtained using the "rigid" and "elastic" hypothesis, there is also a little difference between the two cases for these same angles (180° and 540°).

Figure 4.24 shows the pressure and thickness fields obtained at crank shaft angle 0° at the level of bearing no. 5 for an elastic calculation and a rigid one. Very significant differences can be observed on the geometric profile of the areas of contact and pressure distribution. The pressure fields produce a reactive moment on each bearing. For the case shown in Figure 4.24, the reactive moment changes sign when the elasticity of the bearing is not taken into account.

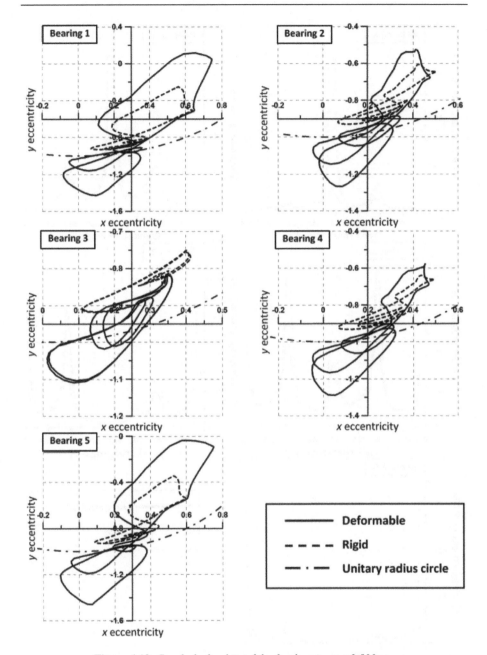

Figure 4.19. *Crank shaft orbits of the five bearings at 2,500 rpm*

184 Internal Combustion Engine Bearings Lubrication in Hydrodynamic Bearings

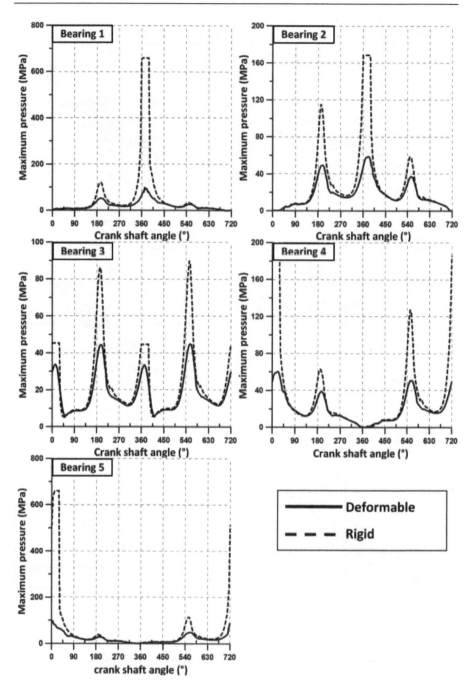

Figure 4.20. *Variation of the maximum pressure at 2,500 rpm*

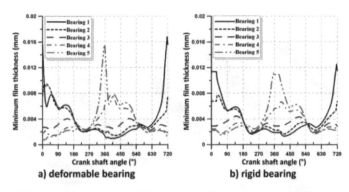

Figure 4.21. *Variation of the minimum film thickness at 2,500 rpm*

Figure 4.25 shows the loads calculated for the five main bearings. The elastic/rigid comparison shows the significant differences on the maximum loads borne by each bearing. In particular, rapid variations in load can be observed for the rigid case.

Table 4.3 summarizes the main results obtained for 2,500 rpm. The comparison between the two calculations shows that the simulation carried out without taking the local elasticity of the bearings into account leads to slightly lower oil flow rates and almost equal rotational power dissipation. Given that the computing time is between 4 and 7 times greater if elasticity is taken into account, a rigid calculation can be considered as a good compromise for an initial approximation of the crank shaft's behavior. However, if the dimension criteria for the bearings are based on maximum pressure or minimum thickness, an elastic calculation is necessary.

	Deformable case					Rigid case				
Crank shaft bearings	1	2	3	4	5	1	2	3	4	5
Rot. Power dissipation[3] (W)	56.8	42.1	40.1	43.5	53.3	56.6	39.9	39.4	40.9	54.0
Oil flow (dm^3/min)	0.44	0.42	0.42	0.43	0.45	0.37	0.38	0.38	0.38	0.37
Min. film thickness (μm)	0.97	1.43	2.09	1.44	0.75	1.29	1.64	1.89	1.68	1.49
Maximum pressure (MPa)	98.8	58.9	45.0	60.2	96.7	660	168	89.8	188	660

Table 4.4. *Main results at 2,500 rpm*

---

3 The rotational power dissipation is obtained by multiplying the friction torque exerted by the shearing of the lubricant and the contact friction stress by the angular rotational speed. From the part coming from the lubricant, it differs slightly from the dissipated power obtained by integrating the dissipation energy with the volume occupied by the lubricant.

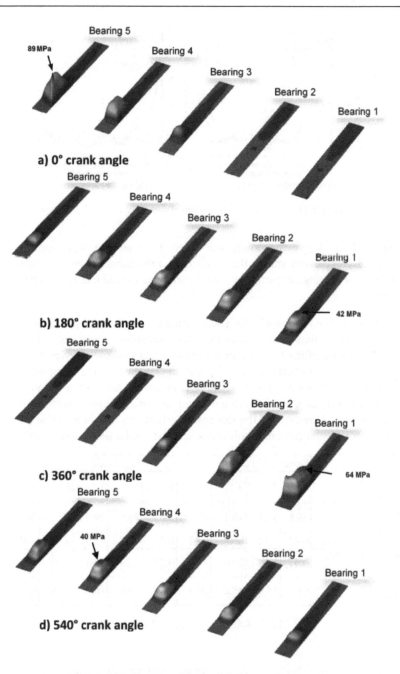

Figure 4.22. *Pressure fields at 2,500 rpm. Elastic case*

The Engine Block–Crank Shaft Link 187

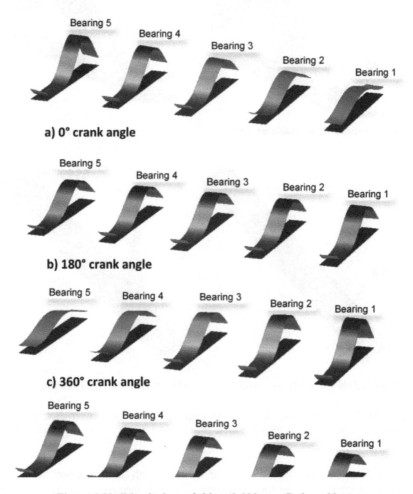

Figure 4.23. *Film thickness fields at 2,500 rpm. Deformable case*

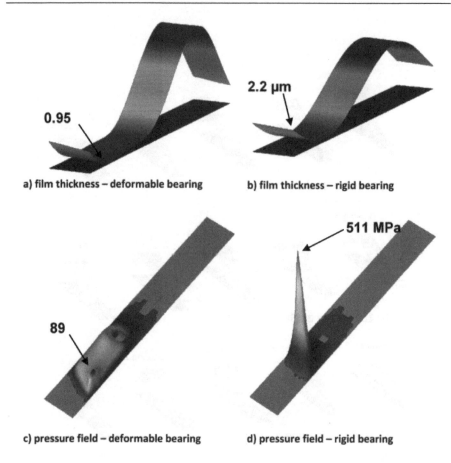

Figure 4.24. *Film thickness [a) and b)] and pressure [c) and d)] fields at 0° crank angle, for bearing no. 5, obtained with deformable case [a) and c)] and rigid case [b) and d)]*

Calculations at 4,000 rpm are carried out with and without taking into consideration the centrifugal forces associated with different parts of the crank shaft and in particular those of the balancing masses. Figure 4.26 shows the orbits of the centers of the five journal bearings. In a general sense, the centrifugal forces slightly increase the eccentricity of the bearings. However, these forces are almost negligible in the significant eccentricity areas with low thickness levels.

The Engine Block–Crank Shaft Link 189

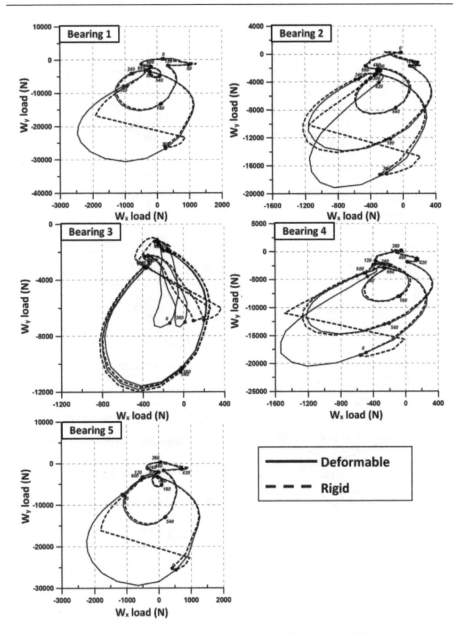

Figure 4.25. *Efforts computed for each main bearing at 2,500 rpm*

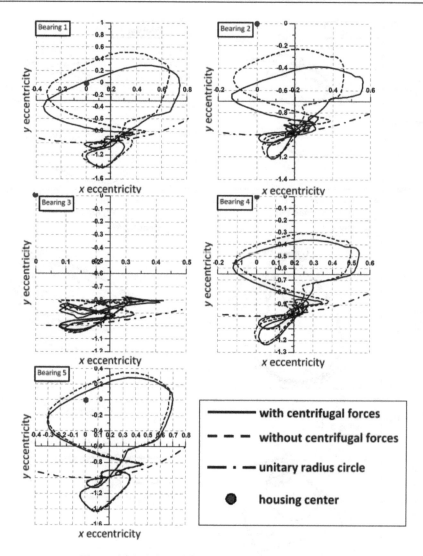

Figure 4.26. *Orbits of the shaft in bearings at 4,000 rpm*

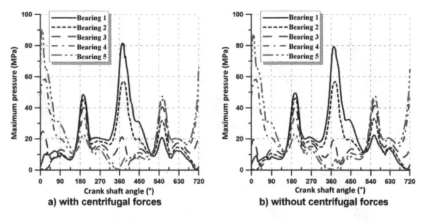

Figure 4.27. *Variation of the maximum pressure at 4,000 rpm*

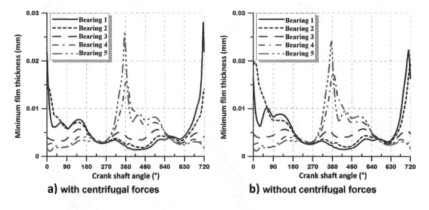

Figure 4.28. *Variation of the minimum film thickness at 4,000 rpm*

Figure 4.27 shows the variation in maximum pressure depending on the angle of the crank shaft. It can be observed that whether the centrifugal forces are taken into account or not has little influence on the maximum pressure values. Figure 4.28 shows the variation in minimum thickness depending on the crank shaft angle. Like for the maximum pressure, whether or not centrifugal forces are taken into account does not have a significant effect on the minimum thickness values, especially when this value is low.

Figure 4.29 shows the pressure fields for the five journal bearings obtained at the crank shaft angles, which correspond to the ignition angles. Like for the 2,500 example, the two bearings under the most severe conditions are bearings no. 1 and no. 5.

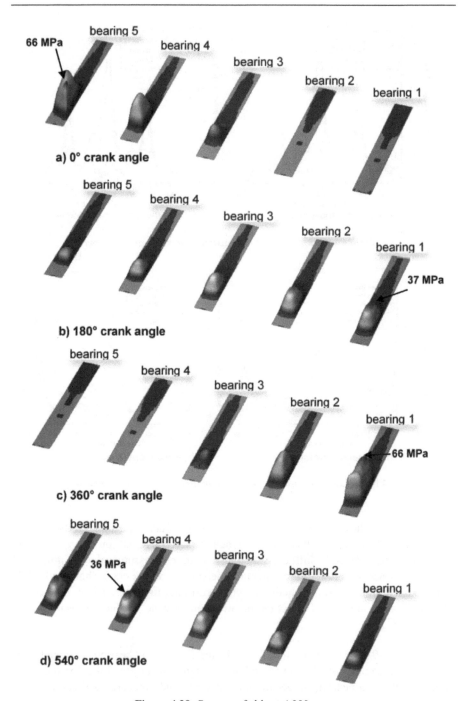

Figure 4.29. *Pressure fields at 4,000 rpm*

Table 4.5 summarizes the main results obtained for 4,000 rpm. At this speed, and even more so for lower speeds, whether centrifugal forces are taken into account or not has a very minor effect on the results.

	With centrifugal forces					Without centrifugal forces				
Main bearings	1	2	3	4	5	1	2	3	4	5
Rot. Power dissipation (W)	95.3	84.8	83.3	87.1	94.5	94.6	85.2	82.2	87.1	94.8
Oil flow (dm^3/min)	0.46	0.43	0.43	0.44	0.46	0.46	0.43	0.42	0.44	0.47
Min. film thickness (µm)	1.38	1.83	2.78	1.86	1.01	1.42	1.93	2.63	1.87	1.11
Maximum pressure (MPa)	81.7	57.0	40.6	58.4	90.2	79.4	56.9	41.4	58.3	86.6

Table 4.5. *Main results at 4,000 rpm*

The calculations for 6,000 rpm are also carried out with and without taking centrifugal forces exerted on various parts of the crank shaft.

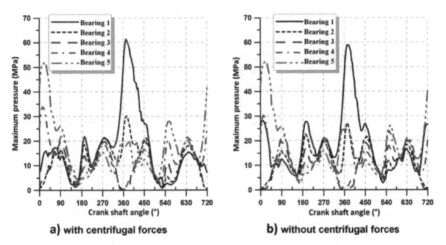

a) with centrifugal forces   b) without centrifugal forces

Figure 4.30. *Variation of the maximum pressure at 6,000 rpm*

Figure 4.30 shows the variation in maximum pressure depending on the crank shaft angle. The differences between the evolutions obtained with and without centrifugal forces start to become more significant. Figure 4.31 shows the variation in minimal thickness depending on the crank shaft angle. Whether centrifugal forces are taken into account or not leads to clear differences in the results. For instance, for bearing no. 1, the differences can reach up to around 10 µm.

Figure 4.32 shows the loads calculated on the five main bearings. Except for bearing no. 5, the comparison between the calculations with and without considering centrifugal effects can differ by up to 1,000 N for the loads borne by the bearings.

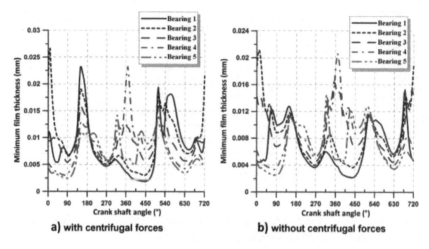

Figure 4.31. *Variation of the minimum film thickness at 6,000 rpm*

Table 4.6 summarizes the main results obtained for 6,000 rpm. The differences in minimum thickness over a whole cycle are around 1 μm for the central bearing while it was only 0.15 μm at 4,000 rpm (Table 4.5). This is due to more or less significant bending of the crank shaft resulting from the differences in the loads exerted on the bearings depending on whether centrifugal forces are taken into consideration or not (Figure 4.32).

	With centrifugal forces					Without centrifugal forces				
Main bearings	1	2	3	4	5	1	2	3	4	5
Rot. Power dissipation (W)	160.6	141.6	132.2	145.9	161.8	161.6	142.7	133.5	145.6	162.0
Oil flow (dm^3/min)	0.75	0.69	0.68	0.71	0.77	0.76	0.71	0.68	0.72	0.77
Min. film thickness (μm)	1.88	3.01	5.38	2.99	2.46	2.16	3.66	4.47	3.34	2.52
Maximum pressure (MPa)	61.4	30.2	20.4	33.9	51.9	59.0	27.2	20.2	30.4	52.4

Table 4.6. *Main results at 6,000 rpm*

At the speed of 6,000 rpm, and even more so for higher speeds and/or more flexible crank shafts (like those used in formula 1), it is essential to take centrifugal forces into account in order to obtain correct data on the evolution of minimum lubricant film thickness and a correct localization of possible contact areas between the crank shaft and the bearing sleeves.

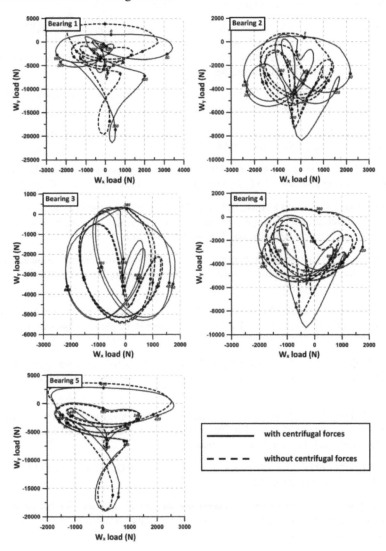

Figure 4.32. *Efforts computed for each main bearing at 6,000 rpm*

4.5. Bibliography

[BON 99] BONNEAU D., CHOMAT-DELALEX A.M., GARNIER T., *et al.*, "Influence of the engine block/crankshaft design on the 3D EHD lubrication of a four cylinder inline automotive engine", *Proceedings of the 10th World Congress on Theory of Machines and Mechanisms,* University of Oulu, Finland, pp. 2488–2495, 1999.

[BON 00] BONNEAU D., CHOMAT-DELALEX A.M., GARNIER T., *et al.*, "3D EHD lubrication optimised design of a four cylinder in line automotive engine crankshaft", *Proceedings of the 26th Leeds-Lyon Symposium., Thinning Films and Tribological Interfaces,* D. Dowson Editor, Elsevier (ed.), Amsterdam, pp. 391–398, 2000.

[BON 14a] BONNEAU D., FATU A., SOUCHET D., *Mixed Lubrication Hydrodynamic Bearings,* ISTE, London and John Wiley & Sons, New York, 2014.

[BON 14b] BONNEAU D., FATU A., Souchet D., *Hydrodynamic Bearings,* ISTE, London and John Wiley & Sons, New York, 2014.

[BON 14c] BONNEAU D., FATU A., SOUCHET D., *Thermo-hydrodynamic Lubrication in Hydrodynamic Bearings,* ISTE, London, and John Wiley & Sons, New York, 2014.

[CHO 01] CHOMAT-DELALEX A.M., BONNEAU D., "Modélisation par la méthode des éléments finis de la lubrification EHD des paliers de vilebrequin des moteurs thermiques", *Revue européenne des éléments finis,* vol. 10, pp. 791–814, 2001.

[FAT 05a] FATU A., Modélisation numérique et expérimentale de la lubrification de paliers de moteur soumis à des conditions sévères de fonctionnement, Doctorate Thesis, University of Poitiers, France, 2005.

[FAT 05b] FATU A., HAJJAM M., BONNEAU D., "An EHD model to predict the interdependent behavior of two dynamically loaded hybrid journal bearings", *Journal of Tribology,* vol. 127, pp. 416–424, France, 2005.

[GAR 97] GARNIER T., Etude élastohydrodynamique de la liaison carter- vilebrequin d'un moteur thermique à quatre cylindres en ligne, Doctorate Thesis, University of Poitiers, 1997.

[GAR 99] GARNIER T., BONNEAU D., GRENTE C., "Three-dimensional EHD behavior of the engine block/crankshaft assembly for a four-cylinder inline automotive engine", *Journal of Tribology,* vol. 121, pp. 721–730, 1999.

[HOD 74] HODGETTS, The vibrations of a crankshaft, PhD Thesis, Cranfield Institute of Technology, Bedford, UK, 1974.

[MOR 01] MOREAU H., Mesures des épaisseurs du film d'huile dans les paliers de moteur automobile et comparaisons avec les résultats théoriques, Doctorate Thesis, University of Poitiers, France, 2001.

[MOR 02] MOREAU H., MASPEYROT P., CHOMAT-DELALEX A.M., *et al.*, "Dynamic behavior of elastic engine bearings – theory and experiments", *Journal of Engineering Tribology,* vol. 216, pp. 179–193, 2002.

5

Influence of Input Parameters and Optimization

Although many numerical and experimental studies about bearings of internal combustion engines show the importance of considering certain phenomena in the modeling phase, such as elastic deformation of surfaces, inertial deformation, thermoviscosity and piezoviscosity of the oil etc., the bibliographic study carried out by [LAV 12] shows that little work has been done on the influence of input parameters (cleanliness of the lubricant, supply pressure and temperature, etc.) on the behavior of the bearing and the optimization of these parameters.

This chapter describes a simple, quick and reliable approach to optimize a number of areas related to connecting rod big end bearings. This approach relies on the design of experiment methods, used to create mathematical models of the dissipated power and functioning severity. These models are then used to replace the cumbersome numerical bearing simulations, during the optimization phase. A large part of this chapter's contents is taken from the thesis of Thomas Lavie, entitled "Optimisation de la Lubrification des Paliers de Tête de Bielle: Démarche"[1] [LAV 12].

5.1. Design of experiments method

The design of experiments (DOE) method or technique is a systematic, rigorous and ordered approach based on statistical considerations, which allows the study of relations between the input and output parameters of a system. It is widely used in a number of areas, such as the agro-food and chemical sectors, but is less

---

[1] *"Optimization of Connecting Rod Big End Bearing Lubrication: Methodological Approach"..*

commonly used in mechanics. In practice, the DOE method can be used as long as we are studying the quantity of interest (measured or calculated) y depending on input variables x_i:

$$y = f(x_i) \qquad [5.1]$$

with $i = 1, ... N$; N showing the number of input parameters.

In the majority of applications, an experiment (also called a "test") is expensive. This is due to the means to be implemented in order to carry out an experiment, such as the time spent by the experiment supervisor, the calculation time for a numerical simulation, etc. The DOE method helps to optimize the selection and the number of experiments to be carried out in order to eventually obtain the maximum amount of information and the greatest level of accuracy possible on the results.

In the context of DOE, the quantity of interest is known as the "response" [GOU 97]. This is the quantity measured at each test by the experiment supervisor. It can also be the result of a calculation in the context of a numerical study. The response generally depends on several variables known as "factors". These are the input parameters of the system. The modification of the factor values can lead to a different response being measured or calculated.

The value given to a factor to carry out a test is called "level" or "modality", and its variation range is called the "factor domain". This can be represented by a graduated and oriented axis (Figure 5.1).

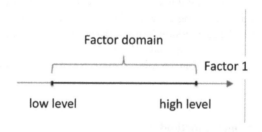

Figure 5.1. *Factor domain*

To take a second factor into consideration, this can be represented in the same way by a second axis which is orthogonal to the first. The area created by these two axes is called the "experimental space", and the zone where the factor domains meet is used to establish the "study domain" (Figure 5.2). This is the area where the experiment supervisor will carry out the tests.

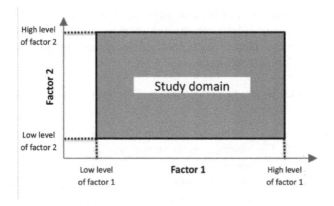

Figure 5.2. *Study domain for two factors*

In the DOE method, most interest should be attached to the relation between a response *y* and *N* factors which, in practice, takes the form of a polynomial:

$$y = a_0 + \underbrace{\sum_{i=1}^{N} a_i x_i}_{first\ order} + \underbrace{\sum_{i=1;j>1}^{N} a_{ij} x_i x_j}_{\substack{interaction\ between \\ two\ factors}} + \underbrace{\sum_{i=1}^{N} a_{ii} x_i^2}_{second\ order} + \cdots \quad [5.2]$$

with x_i as the level of factor *i*, *y* is the value of the response, and a_0, a_i, a_{ij}, a_{ii}…are the unknown coefficients that need to be determined using the experimental results.

This is one of the key points in the DOE process. It consists of optimizing the selection of experiments to be carried out in order to build the most representative mathematical model possible of the phenomenon being studied. This model can then be used to predict new response values without having to carry out any further experiments.

The number of experiments to be carried out depends directly on the complexity of the mathematical model chosen by the experiment supervisor. Two firm hypotheses are usually used:

– *Hypothesis 1*: the interactions of more than two factors are negligible, which means that only the interactions x_i and x_j are taken into account by the model.

– *Hypothesis 2*: within the variation range, the responses can be correctly approximated by second-degree polynomials.

There are studies for which these hypotheses are not valid. It is, therefore, possible to use higher order mathematical models, which take into account the interactions of more than two factors. Nevertheless, in practice, these two hypotheses work for most applications [GOU 97].

The different DOEs can be divided into two categories according to the objective of the experiment:

– Screening method: helps to determine the factors which influence the response observed. The objective is to answer yes or no to the question – does parameter x_i have an influence? The goal is to minimize the problem by eliminating the factors that have little or no influence from the study. For this method, a large number of factors can be studied with few tests. The mathematical models for screening are first-order polynomials, with or without the consideration of the two factor interactions:

- Model without interaction:

$$y = a_0 + \sum_{i=1}^{N} a_i x_i \qquad [5.3]$$

- Model with interactions:

$$y = a_0 + \sum_{i=1}^{N} a_i x_i + \sum_{i=1, j>i}^{N} a_{ij} x_i x_j \qquad [5.4]$$

– *Optimization* method: is used to produce a predictive mathematical model of the system. The objective is to model the behavior of the response studied with better accuracy. This method is used during the optimization phase. This method requires more experiments than the screening method, for the same number of factors. For optimization, the mathematical models are generally second-degree polynomials, which consider the interactions of the two factors:

$$y = a_0 + \sum_{i=1}^{N} a_i x_i + \sum_{i=1, j>i}^{N} a_{ij} x_i x_j + \sum_{i=1}^{N} a_{ii} x_i^2 \qquad [5.5]$$

A detailed description of the two categories and also the various statistical tools used to verify the validity and accuracy of the models obtained are given in [LAV 12].

5.2. Identification of the input parameters: example

The first study based on the DOE method to determine the most influential input parameters on the behavior of a connecting rod big end bearing was published in 2009 [FRA 09]. The influence of 10 factors (Table 5.1) was studied with respect to three global parameters (dissipated power, leakage flow rate and operating temperature) and two local parameters (minimum film thickness and maximum film pressure). A screening method is used, more precisely a fractional factorial design, which reduces the number of simulations to $2^5 = 32$. The mathematical models determined cannot, therefore, be used to accurately predict the behavior of the bearing for a possible optimization. However, it has been proven that the five most influential factors are, in this order: the viscosity of the lubricant, the length of the bearing, the thermoviscosity coefficient, the speed of the engine and the radial clearance.

Factor name	Notation	Unit	Low level	High level
Engine speed		rpm	4,500	6,000
Viscosity	VIS	Pa.s	0.01	0.05
Oil thermo-viscosity coefficient	THV	°C	0.01	0.05
Oil piezo-viscosity coefficient	PIV	MPa^{-1}	0.005	0.02
Bearing length	LEN	mm	16	20
Bearing clearance	CLE	µm	10	50
Lemon shape	LEM	µm	0	10
Barrel shape	BAR	µm	0	2
Supply circumferential position	LOC	°	30	60
Supply pressure	PSU	MPa	0.2	1

Table 5.1. *Factor values for the screening plan*

Figure 5.3 shows the effect of the factors on the responses examined. It can be observed, for instance, that although the radial clearance (CLE) is the most influential factor on flow rate, minimum film thickness and operating temperature, it does not have a particularly significant role on power loss, and influences the maximum pressure less than the engine speed and the lemon shape of the bearing housing. We can also conclude that an increase in the clearance would lead to an increase in the flow rate, the maximum pressure and minimum film thickness, but to a decrease in the operating temperature.

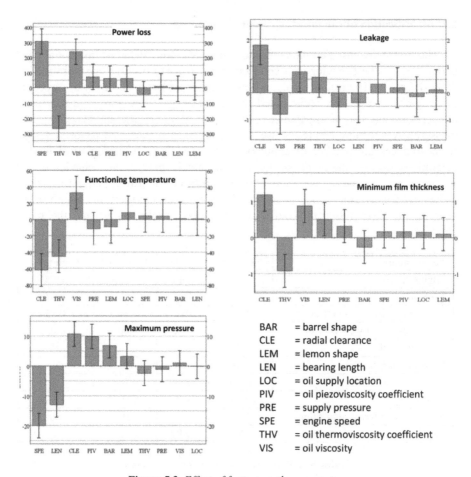

Figure 5.3. *Effect of factors on the parameters*

5.3. Multiobjective optimization

For most industrial applications, a system to be optimized is characterized by several responses. These responses, also called "objectives" in the context of an optimization, help to assess the quality of the system's operation. Multiobjective optimization consists of simultaneously optimizing the responses which are, in general, contradictory.

For instance, let us take the simplified case of an internal combustion engine only considering two objectives to optimize: performance and consumption. The objective is to increase the engine's performance while also decreasing its consumption. It is clear that it is impossible to obtain the performance levels of a sports car engine with the consumption of a small economical car engine. The two objectives are contradictory; an increase in one objective generally leads to the deterioration of the other. For a multiobjective optimization, there is not one single solution, which optimizes all objectives, but rather, a number of compromises known as "Pareto optimality" (Figure 5.4). The resolution of a multiobjective solution, therefore, consists of determining one or several optimal solutions.

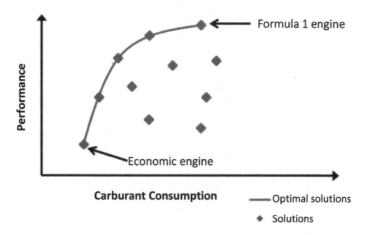

Figure 5.4. *Illustration of a multiobjective optimization problem*

The advantages and disadvantages of the various optimization methods are presented in detail in section "Multi-objective optimization" of Lavie's thesis [LAV 12]. For a study dedicated to the optimization of a connecting rod bearing, the method chosen is based on evolutionary algorithms (EAs).

The EAs or genetic algorithms are inspired by Darwin's theory of evolution, according to which only the individuals who are most well-adapted to their environment will survive and reproduce, leading to an even better adaptation of future generations to their environment. The EAs reproduce this concept of species' evolution in the context of a multiobjective optimization problem.

The basis of the EAs is given in Figure 5.5:

– The first parent population random assortment.

– The best individuals from the parent population are then selected.

– Crossing and mutation are used to create a child population.

– The parent population is replaced by the child population (start of a new cycle).

The EAs enable the approximation of the entirety of the Pareto surface of a multiobjective problem with a uniform distribution of solutions, without leading research with arbitrary parameters such as weight or constraints. The main disadvantage of EAs is the high number of calculations carried out to reach the true Pareto surface of a problem.

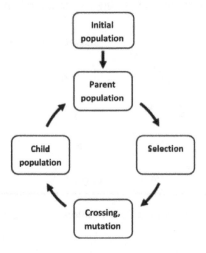

Figure 5.5. *Base algorithm of an evolutionary algorithm* [LAV 12]

The optimization methodology developed in [LAV 12] consists of using regression models obtained by the DOE method for optimization by EA. All of the solutions to the problem can then be obtained almost instantaneously. In the various optimizations carried out, two objectives are considered: the dissipated power *POW* and a criterion for severity (*CPV* or *SEV*). The goal is to minimize the dissipated power while also reducing the harshness of the contact.

The criterion of severity *SEV* represents the percentage of the bearing surface, which works during the cycle below a certain film thickness fixed by the user. It can give an indication on the proportion of the surface for which it occurs important contact risks between surfaces. More the high value of *SEV,* more the surface function in severe conditions. *SEV* strongly related to the contact pressure Pc, since it is calculated starting from the thickness of film [BON 14a].

5.4. Optimization of a connecting rod big end bearing: example

The study presented in this chapter was carried out on a diesel engine connecting rod. The characteristics of the connecting rod are given in Table 5.2. The film mesh used consists of four elements over half the width of the bearing and 64 elements around the circumference (Figure 5.6). The estimates of calculation times for isothermal (*ISO*) calculations and global thermal (*GT*) calculations (see section 1.1 of [BON 14b]) are given in Table 5.3. The ISO model is around 5 times faster than a GT calculation. The calculation times are high but are compatible with the studies to be carried out, and the mesh seems to be a good compromise between the calculation time and the accuracy of the representation. A study on the influence of the mesh is presented in section "Study of meshing influence" of Lavie's thesis [LAV 12].

	Global thermal model	**Isothermal model**
Computation time	210	40

Table 5.2. *Characteristics of the studied connecting rod*

Bearing diameter	47.26	mm
Bearing length	16.7	mm
Crank shaft radius	44	mm
Connecting rod length	137	mm
Piston diameter	88	mm
Connecting rod mass	0.6	kg
Piston mass	0.7	kg
Connecting rod inertia	2,000	kg.mm^2
Axial supply length	8	mm
Supply circumferential position	20	°

Table 5.3. *Estimation of the computing time*

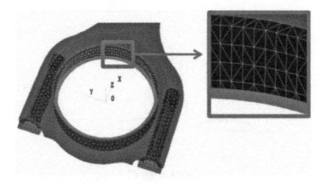

Figure 5.6. *3D mesh of a connecting rod big end bearing. For a color version of the figure, see www.iste.co.uk/bonneau/hydrobearings4.zip*

Four operating conditions were studied (speed and load):

– 2,000 rpm at 100 % of the load (Figure 5.7);

– 2,000 rpm at 30 % of the load (Figure 5.8);

– 4,000 rpm at 100 % of the load (Figure 5.9);

– 4,000 rpm at 30 % of the load (Figure 5.10).

The full load operating conditions correspond to a driver applying maximum acceleration while going uphill. These are the harshest conditions applied to the connecting rod big end bearings. In this chapter, the operating condition will be noted as follows: "speed"_ "load percentage".

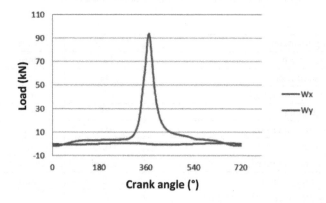

Figure 5.7. *Load at 2,000 rpm and 100 % of the maximum load (case 2000_100). For a color version of the figure, see www.iste.co.uk/bonneau/hydrobearings4.zip*

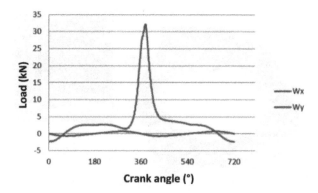

Figure 5.8. *Load at 2,000 rpm and 30 % of the maximum load (case 2000_30). For a color version of the figure, see www.iste.co.uk/bonneau/hydrobearings4.zip*

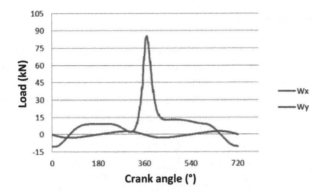

Figure 5.9. *Load at 4,000 rpm and 100 % of the maximum load (case 4000_100). For a color version of the figure, see www.iste.co.uk/bonneau/hydrobearings4.zip*

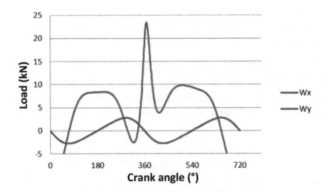

Figure 5.10. *Load at 4,000 rpm and 30 % of the maximum load (case 4000_100). For a color version of the figure, see www.iste.co.uk/bonneau/hydrobearings4.zip*

The responses chosen for the optimization are the dissipated power (*POW*) and a severity criterion (*CPV* or *SEV*). The goal is to minimize the dissipated power – that is the energy lost through friction – while ensuring the longevity and reliability of the connecting rod.

The choice of factors' variation ranges was carried out on parameters, which seemed beforehand to be influential on the behavior of the bearing, and which are controlled to a limited extent or are poorly controlled. The variation ranges are determined from existing uncertainties. The optimization of the bearing shows the improvement that can be made with a better mastery of these factors.

5.4.1. Viscosity factors

Measurements of the viscosity of two oils 5W30 of different brands were carried out with a falling sphere viscometer over a large temperature and pressure range (30–150°C and from 0 to 400 MPa). The coefficients of the thermoviscosity[2] and piezoviscosity[3] laws are given in Table 5.4. The reference temperature is fixed at 80°C.

	Unit	Oil A	Oil B
μ_0	Pa.s	0.0132	0.0137
μ_s	Pa.s	0.0021	0.002
β	°C^{-1}	0.0273	0.0277
a	MPa^{-1}	0.00212	0.0021
b		6.35	6.16

Table 5.4. *Measured thermoviscosity an d piezoviscosity coefficients*

The high temperature viscosity grade of Society of Automotive Engineers (SAE) standard 30 means that the kinematic viscosity $v = \frac{\mu}{\rho}$ à 100 °C is between 9.3 and 12.5 centistokes. In order to determine the kinematic viscosities, the density of the two oils was measured at 100°C with the help of a pycnometer. Table 5.5 shows the results of measurements, as well as the kinematic viscosities at 100°C. The two oils have very similar thermoviscous behavior, which is near the upper limit of the SAE classification.

	Unit	Oil A	Oil B
ρ	kg.m^{-3}	789	790
μ at 100 °C	Pa.s	0.0097	0.0098
v at 100 °C	cSt	12.36	12.5

Table 5.5. *Density and kinematic viscosity at 100°C*

The domains of factors μ_0, μ_s and β were chosen, so that they contain the upper and lower limits on the high temperature viscosity for grade 30 of the SAE standard (Table 5.6). Coefficients μ_0 and μ_s are combined to form a single factor named "*VIS*". *VIS* represents the percentage variation of μ_0 and μ_s around their central values ($\mu_0 = 0.0125$ Pa.s and $\mu_s = 0.002$ Pa.s). Figure 5.11 shows the

---

2 $\mu = \mu_0 e^{-\beta(T-T_0)} + \mu_s$.

3 $\mu = \mu_0 (1+ap)^b$.

thermoviscosity domain covered by the design of experiment plan. The domains for a and b were arbitrarily established using measured values.

	Factor name	Unit	Level −1	Level 0	Level +1
$\Delta\mu$	VIS	%	−16	0	16
β	BET	°C^{-1}	0.0252	0.028	0.0308
a	COA	MPa^{-1}	0.0018	0.002	0.0022
b	COB		5.8	6.3	6.8

Table 5.6. *Viscosity factor domains*

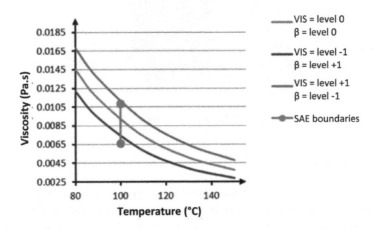

Figure 5.11. *Thermoviscosity domain covered by the design of experiment plan. For a color version of the figure, see www.iste.co.uk/bonneau/hydrobearings4.zip*

5.4.2. *Radial clearance factor*

The clearance factor (*CLE*) represents the radial clearance of the bearing. It is between 12 and 40 μm, and its domain was defined using the manufacturer's tolerance levels.

5.4.3. *Radial shape defect*

The "lemon shape" factor (*LEM*) characterizes a non-circular form of the bearing: the radial clearance varies depending on the circumference of the bearing. Lemon shape measurements were carried out on several connecting rods for different screw tightening torques. The lemon shape values measured are lower than 4 μm. The lemon shape factor domain was set at between 0 and 4 μm.

5.4.4. *Axial shape defect*

The "barrel shape" factor (*BAR*) characterizes the barrel shape of the connecting rod. The radial clearance of the connecting rod varies depending on its thickness. The barrel shape may appear during the breaking-in phase, due to wear and/or the matting of the edges of the connecting rod, which aligns itself inside the bearing. Measurements were carried out on several connecting rods following normal usage. The barrel shape values measured are below 2 µm. The barrel shape factor domain is set at between 0 and 2 µm.

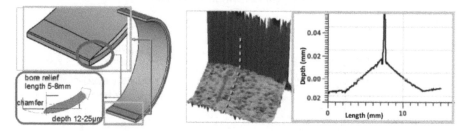

Figure 5.12. *Shell bore relief factors*

5.4.5. *Shell bore relief factors*

Two factors linked to shell bore relief were taken into account (Figure 5.12): length (*LEN*) and depth (*DEP*). The domains of these factors were defined using the manufacturer's tolerance levels: the length is between 5 and 8 mm, and the depth is between 12 and 25 µm.

5.4.6. *Supply pressure and temperature*

The supply pressure (*PSU*) and temperature (*TSU*) domains were arbitrarily defined using actual operating values. The pressure is between 0.1 and 0.5 MPa and the temperature is between 90 and 120°C.

The domains of all factors are given in Table 5.7. The first four factors (*VIS*, β, *a* and *b*) are connected to the rheological properties of the oil. The factors *CLE*, *LEM*, *BAR*, *LEN* and *DEP* are connected to the geometric shape of the connecting rod. Finally, the factors *TSU* and *PSU* are connected to the conditions of oil supply to the bearing.

In Chapter 3 of [LAV 12], two studies are carried out: the first (study no. 1) deals only with factors *VIS*, *BET*, *COA*, *COB*, CLE and *TSU*, whereas the second (study no. 2) looks at all 11 factors described above. Study no. 1 helped to validate the optimization methodology while avoiding having to carry out too many calculations for EA optimization by directly using the simulation program. Moreover, since there are fewer factors than in study no. 2, the mathematical models obtained are more accurate. Study no. 2 was then carried out to create a more ambitious optimization of the bearing (with a higher number of factors). Some aspects of this second study are discussed in the following pages.

Factor name	Notation	Level −1	Level 0	Level +1
Viscosity (%)	VIS	-16	0	16
β (°C)	BET	0.0252	0.028	0.0308
a (MPa^{-1})	COA	0.0018	0.002	0.0022
b	COB	5.8	6.3	6.8
Clearance (µm)	CLE	0.012	0.026	0.040
Lemon shape (µm)	LEM	0	0.002	0.004
Barrel shape (µm)	BAR	0	0.001	0.002
Length (mm)	LEN	5.0	6.5	8.0
Depth (mm)	DEP	0.012	0.0185	0.025
Supply temperature (°C)	TSU	90	105	120
Supply pressure (MPa)	PSU	0.1	0.3	0.5

Table 5.7. *Factors specification*

5.4.7. *Power loss*

The influence of the factors on power loss response *POW* is shown in Figure 5.13. It can be observed that a reduction in *BAR* and *PSU* leads to a fall in *POW*. *LEM* has little or no influence. The factors linked to shell surface relief have no influence.

5.4.8. *Contact pressure velocity factor*

The influence of factors on the product response *CPV* is shown in Figure 5.14. *BAR* has a very significant influence on *CPV*. An increase in *BAR* leads to a reduction in *CPV*. This can be explained by the fact that the lubrication is mixed at the edges of the bearing. The barrel shape induces an increase in the radial clearance at the edges of the bearing, which leads to a reduction in P_cV. However, if the barrel shape is too pronounced, the lubrication can become of the mixed regime at the center of the bearing. Therefore the sleeve must have the capacity to adapt itself to this regime during the breaking-in phase. A significant supply pressure can reduce the *CPV* product. Factors *LEM*, *LEN* and *DEP* have no influence.

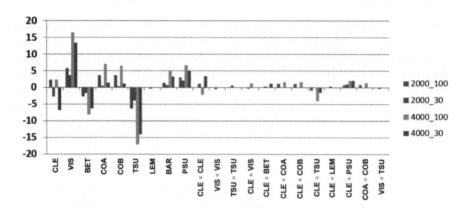

Figure 5.13. *Factor influence on the power loss (response POW). For a color version of the figure, see www.iste.co.uk/bonneau/hydrobearings4.zip*

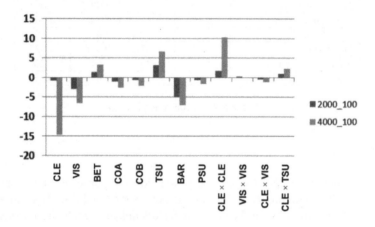

Figure 5.14. *Factor influence on response CPV. For a color version of the figure, see www.iste.co.uk/bonneau/hydrobearings4.zip*

5.4.9. *Severity criterion based on the minimum film thickness*

The coefficients of the models for *SEV* response are shown in Figure 5.15. Unlike *CPV*, *BAR* does not have a significant influence on *SEV*. The barrel shape can, therefore, help to reduce *CPV*, which mainly appears at the edges of the bearing but does not really have an influence on the overall severity criterion *SEV*. The modification of *LEM* does not have a significant influence on *SEV*, likewise for *PSU*. The latter does not appear to be a parameter to choose to improve the performance of the bearing, contrary to what we might expect. This conclusion is confirmed by tests carried out by a manufacturer, which demonstrated that with low values of *PSU*, the bearing remains functional. Factors *LEN* and *DEP* have no influence.

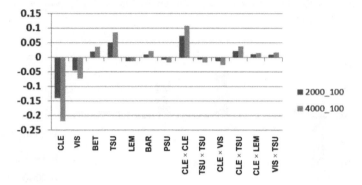

Figure 5.15. *Factor influence on the severity criterion based on the minimum film thickness (response SEV) For a color version of the figure, see www.iste.co.uk/bonneau/hydrobearings4.zip*

5.4.10. *Leakage*

The models for the leakage response (notated *FLO*) are shown in Figure 5.16 for all operational conditions. The factors *BAR* and *LEN* have no influence on the leakage flow rate, which is not surprising given that most of the leakage is situated in the high film thickness areas, which are not influenced by deformation. Factors *DEP* and *LEM* have no influence on their own, but the interaction between the two does have an effect, most probably due to the accumulation, which significantly modifies the clearance at the shell chamfers near the bore plane. There is a very significant interaction between the *CLE* and the *PSU*.

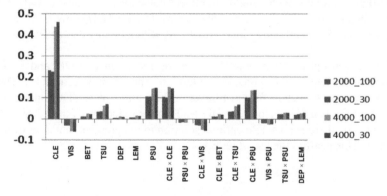

Figure 5.16. *Factor influence on the leakage (response FLO). For a color version of the figure, see www.iste.co.uk/bonneau/hydrobearings4.zip*

5.4.11. *Global functioning temperature*

The models for *TEM* response are shown in Figure 5.17. Only, *CLE* and *TSU* have any real influence.

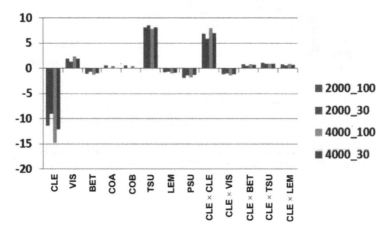

Figure 5.17. *Factor influence on the functioning temperature (response TEM). For a color version of the figure, see www.iste.co.uk/bonneau/hydrobearings4.zip*

5.4.12. *Bearing optimization method*

In order to test the metamodel methodology, it is compared to a traditional optimization procedure directly using the harshest calculation case: 2000_100. It is, therefore, a question of testing the robustness of the method in the least favorable lubrication conditions, with a highly nonlinear behavior of responses. Regarding the results obtained by DOE, it seems unnecessary to take *LEM*, *LEN*, *DEP* and *PSU* into account (*PSU* significantly increases *POW*, without really decreasing *CPV*).

The chosen optimization objectives are the reduction of *POW* and *CPV*. The coefficients of the two responses' models are shown in Figure 5.18. The two responses are contradictory for most factors, where an increase of a response caused by the modification of a factor generally leads to the reduction of the other response. Consequently, there is no single optimal solution to the problem, but a set of optimal solutions. This justifies the use of an EA.

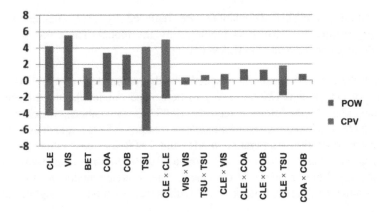

Figure 5.18. *Power loss (POW) and P_cV product (CPV) models for the case 200_100. For a color version of the figure, see www.iste.co.uk/bonneau/hydrobearings4.zip*

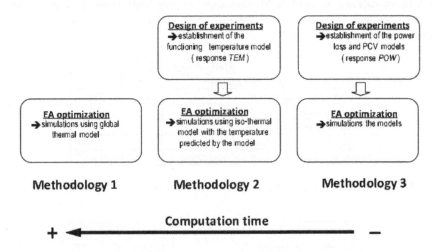

Figure 5.19. *Presentation of the three optimization methodologies*

Two optimization methods are considered:

– The first (Methodology 1 – Figure 5.19) consists of using GT model (*GT*) calculations during EA optimization phase.

– The second (Methodology 2 – Figure 5.19) uses models obtained by DOE directly for optimization.

Given the significant calculation time for a *GT* simulation, a third methodology was suggested (Methodology 3 – Figure 5.19). This consists of combining the model GT for determining bearing operating temperature with *ISO* calculations during the optimization phase. *GT* calculations are replaced by *ISO* calculations at the operating temperature, which is predicted by the model. The three optimization methodologies are presented in [LAV 12].

Tables 5.8 and 5.9 present respectively the parameters of the EAs and the number of calculations (*ISO* and *GT*) to be carried out for the three optimization methodologies. Methodology 1 requires 1,600 *GT* calculations, which equates to around 5,600 h of calculations. Methodology 3 requires 57 *GT* calculations to obtain the bearing operating temperature model and 1,600 *ISO* calculations for the EA optimization phase, equating to around 1,266 h of calculations. Finally, Methodology 2 only requires 57 *GT* calculations to create models for the responses of *POW* and *CPV* (the optimization time with regression models is negligible).

Methodology 3 is around 4 times quicker than Methodology 1, and Methodology 2 is around 6 times quicker than Methodology 3. Given the very significant calculation time of Methodology 1, only Methodologies 2 and 3 were carried out and compared. Nevertheless, some isolated comparisons were made between Methodologies 1 and 2.

Methodology	1 – 3	2
Number of individuals by generation	16	500
Number of generations	100	230
Cross probability	0.9	0.9
Mutation probability	1	1
Total number of individuals	1,600	115,000
Number of values by factor (discretization)	10	100

Table 5.8. *Evolution algorithm parameters for the three optimization methodologies*

Methodology	1	3	2
Number of computations when using the global thermal model	1,600	57	57
Number of computations when using the isothermal model	0	0	1,600
Computation time (hours)	5,600	200	1,266

Table 5.9. *Computation number and computation time estimation for the three optimization methodologies*

Figure 5.20 shows the Pareto fronts obtained with the two methodologies. The results obtained are almost identical. The front obtained with Methodology 2 is slightly wider than the one obtained with Methodology 3, which can be explained by the higher number of individuals (500 compared to 16). It seems clear that

Methodology 2 has significantly more advantages than Methodology 3, since the results are more complete and they can be obtained 6 times quicker.

These results, along with others in section "Methods for optimization of bearing behavior" of Lavie's thesis [LAV 12], show that when metamodels are of a high quality, it can be highly advantageous to use them as replacements for complete calculations. The results obtained by Methodology 2 are reliable from a numerical point of view. It is now necessary to determine the optimal levels of the factors, since this is the ultimate goal of any optimization process.

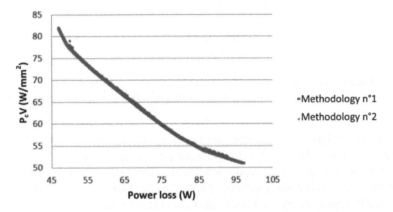

Figure 5.20. *Pareto front comparison for the contact pressure velocity product response. For a color version of the figure, see www.iste.co.uk/bonneau/hydrobearings4.zip*

The responses chosen for optimization are energy loss (*POW*) for the 2000_30 case and the severity criterion on the minimum film thickness *(SEV)* for 2000_100. The 2000_100 case is little used but puts a lot of strain on the bearing. It is more common to find the 2000_30 case, but poses no risk to the bearing.

Figure 5.21 represents the result of the optimization as well as the criterion domain boundaries. In order to determine these boundaries, three new optimizations were carried out, looking at:

– a minimization of *POW* and a maximization of *SEV*;

– a maximization of *POW* and a minimization of *SEV*;

– a maximization of *POW* and a maximization of *SEV*.

The domain boundaries enable the observation of the improvement that can be made for *SEV* for a given value of *POW*. It can be seen that *SEV*, and therefore the severity of the contact, can vary more than three-fold for a given value of *POW*.

218 Internal Combustion Engine Bearings Lubrication in Hydrodynamic Bearings

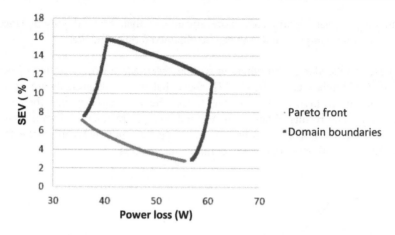

Figure 5.21. *Pareto front and domain boundaries. For a color version of the figure, see www.iste.co.uk/bonneau/hydrobearings4.zip*

The area of interest was chosen arbitrarily by eliminating the optimal solutions with high values of *POW* and with significant severity levels. It is fair to assume that engine manufacturers will generally work within this central zone, which is an area of compromise. This zone is between 38 W < *POW* < 48W. The *SEV* response can be correctly approximated by a linear equation (Figure 5.22):

$$MFT = -0.24 * POW + 15.2 \ + 15.2 \qquad [5.6]$$

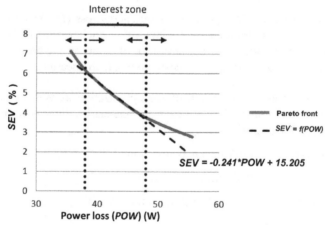

Figure 5.22. *Linear approximation of SEV response in the interest zone. For a color version of the figure, see www.iste.co.uk/bonneau/hydrobearings4.zip*

Figure 5.23 shows the Pareto front (left-hand axis) and the variation of the factors on the Pareto front (right-hand axis). In the chosen interest zone, *VIS*, *BET*, *COA* and *COB* have low values, and *CLE* is at its maximum value. *TSU* varies linearly from its value -1 to 1 such that:

$$TSU = -0.2 * POW + 8.6 \qquad [5.7]$$

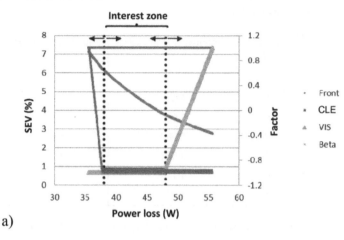

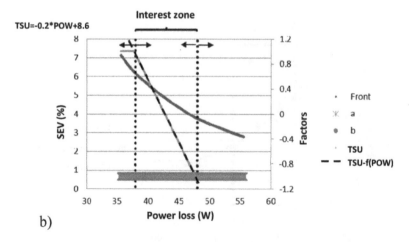

Figure 5.23. *Factor variation on the Pareto front. For a color version of the figure, see www.iste.co.uk/bonneau/hydrobearings4.zip*

Equations [5.6] and [5.7] are used to define six optimal solutions in the area of interest using six required *POW* values. The values of *POW* and *SEV*, as well as those of *TSU*, which are necessary to obtain these solutions, are presented in

Table 5.10. These solutions were calculated in *GT*. The factors *CLE*, VIS, *BET*, *COA* and *COB* are set at their optimal values (high or low level) at this part of the front. The square symbols in Figure 5.24 show the solutions obtained with equation [5.6], and the triangle symbols represent the results of the corresponding *GT* calculations. Given the excellent correlation of the results, it is clear that equations [5.6] and [5.7] can help predict the optimal solutions for the zone of interest, for *SEV* and also for the corresponding values of *TSU*.

Power loss (POW) W	Minimum film thickness (SEV) μm	Supply temperature (TSU) dimensionless
38	6.046	0.95
40	5.565	0.55
42	5.083	0.15
44	4.601	−0.26
46	4.119	−0.66
48	3.637	−1.00

Table 5.10. *Solutions of the Pareto front obtained by GT*

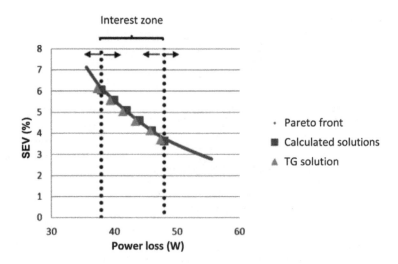

Figure 5.24. *Computation by TG of the six optimal solutions. For a color version of the figure, see www.iste.co.uk/bonneau/hydrobearings4.zip*

With an initial effort required to calculate the metamodels, the behavior of a connecting rod bearing can be perfectly established. As shown by this example, it is then very simple to integrate the behavior of the bearing into a larger code, which deals with, for instance, heat loss or the modeling of bearing seizing, a complex process that involves a number of different parameters [LIG 11].

Two other examples of optimization can be found in section "Optimization of bearing behavior" of Lavie's thesis [LAV 12].

5.5. Bibliography

[BON 14a] BONNEAU D., FATU A., SOUCHET D., *Mixed Lubrication in Hydrodynamic Bearings*, ISTE, London and John Wiley & Sons, New York, 2014.

[BON 14b] BONNEAU D., FATU A., SOUCHET D., *Thermo-hydrodynamic Lubrication in Hydrodynamic Bearings*, ISTE, London and John Wiley & Sons, New York, 2014.

[FRA 09] FRANCISCO A., FATU A., BONNEAU D., "Using design of experiments to analyze the connecting rod big-end bearing behavior", *ASME, Journal of Tribology*, vol. 131, pp. 011101–011113, 2009.

[GOU 97] GOUPY J., "Plans d'expériences", *Techniques de l'ingénieur*, PE 230, p. 26, 1997.

[LAV 12] LAVIE T., Optimisation de la Lubrification des Paliers de Tête de Bielle: Démarche Méthodologique, Doctorate thesis, University of Poitiers, France, 2012.

[LIG 11] LIGIER J.L., DUTFOY L., "Modeling and prediction of a simplified seizure mechanism occurring in conrod bearings", *Mécanique & Industries*, vol. 12, pp. 265–273, 2011.

Index

A, C

ambient pressure, 36, 60, 98, 126, 128, 133, 142, 164, 170
cavitation, 98, 99, 100
   algorithm, 98
compliance matrix, 46, 127, 128, 140, 141, 164, 173–175
computation time, 64, 102, 108, 205, 216
connecting rod
   big end bearing, 24–26, 31–34, 76, 78–80, 82, 88, 89, 101, 102, 104, 109, 110, 113, 116, 118, 161, 166, 197, 201, 204–206
   body, 32, 38–40, 42–44, 46, 52, 56, 58, 82
   small end bearing, 26, 27, 124, 126, 127, 146, 147, 151
contact
   algorithm, 59, 72
   pressure velocity factor, 211
   pressure, 40, 85, 89–91, 97–99, 129, 130, 132, 135, 149, 154, 162, 167–169, 204, 211, 217
crank
   pin, 25, 31, 33, 35–40, 46, 60, 63, 72–79, 84–89, 92, 94, 95, 97, 98, 101–105, 113, 114, 117, 118, 161, 162, 170, 172, 173, 175, 177, 179, 180, 182

shaft angle, 4, 64, 65, 78, 88, 104, 110, 117, 132, 133, 145, 149, 150, 151, 154–157, 165, 166, 168, 182, 191, 193
shaft main bearing, 27–29
shaft radius, 24, 35, 60, 76, 77, 82, 84, 92, 102, 128, 142, 205

D, E

damage, 31, 98–100, 123
design of experiments, 197
diesel engine, 33, 79, 81, 82, 85, 88, 90, 94, 125, 127, 129–131, 140, 163, 166, 204
dissipated
   energy, 153, 155, 156
   power, 105, 107, 109, 112, 113, 117, 185, 197, 201, 204, 207
elastic
   behavior, 33, 38, 172–174, 176, 178
   deformation, 42, 79, 137, 173, 175, 197
elementary solution, 44, 46, 141, 145, 146
engine
   block, 2–4, 8, 11, 27, 28, 31, 161–163, 170–179
   cycle, 8, 27, 76–79, 85, 102, 146, 149, 155, 156, 166, 167
   torque, 28, 29, 142, 146, 147, 151, 176

equilibrium equation, 22, 41, 42, 44, 45, 53, 57, 68, 70, 74, 135–137, 172, 176
evolutionary algorithm, 203, 204

F, G, H, I

factor domain, 198, 209, 210
feeding orifice, 102
film thickness, 40, 58, 62, 74, 85, 92, 93, 95, 96, 101–103, 105, 115, 131–133, 136, 149, 157, 168, 175, 185, 187, 188, 193–195, 201, 204, 212, 213, 217, 220
flow rate, 33, 85, 93–95, 96, 105, 107, 109–113, 117, 185, 201, 213
friction torque, 53, 135, 138, 146, 148, 152, 153, 177, 185
gas pressure, 9, 23, 34, 139, 146, 156, 157, 158
gasoline engine, 33, 79, 80, 81, 85, 89, 101, 102, 116, 117
hydrodynamic pressure, 40, 42, 48, 53, 54, 85, 86, 89–93, 97, 130, 135, 147, 154, 163, 166, 168, 169
inertia deformation, 96
isothermal, 35, 85, 107, 205, 216

J, L, M

Jacobian matrix, 74, 75, 137, 178, 179
joint plane, 39, 40, 42, 43, 46–49, 51, 53, 54, 56, 58, 59, 60, 62–65, 72, 92
leakage, 165, 166, 201, 213
load, 13, 22–29, 31, 34–36, 38, 39, 55, 61, 63, 64, 68, 69, 80–83, 85–92, 101, 106, 115, 118, 124, 128–130, 135–137, 141, 145–148, 151, 154, 157, 165, 167, 180–182, 185, 206, 207
  diagram, 13, 14, 22, 24, 26, 27, 31, 34–36, 61, 63, 64, 80–83, 87, 88, 90, 92, 101, 124
maximum pressure, 23, 34, 36, 38, 76–78, 89, 92, 107, 113, 114, 117, 182, 184, 185, 191, 193, 194, 201
mesh refinement, 96

minimum film thickness, 37, 62, 76–78, 86–88, 96, 97, 102, 103, 105, 107, 109–111, 114, 117, 118, 131, 149, 154, 157, 158, 169, 185, 191, 194, 201, 212, 213, 217, 220
misalignment, 24, 72, 75, 124, 162, 171, 172, 175, 182
mixed lubrication, 8, 40, 43, 130, 148, 151, 167, 176
mobility, 1, 2, 14–16, 18–20, 22, 31, 135
MOFP, 85
MOFT, 85, 86, 88, 110, 113, 115, 117, 149
multi-objective optimization, 203

N, O, P

nodal force, 53, 54, 67
Nusselt number, 106, 107
optimization, 197, 200–204, 207, 211, 214–217, 221
piezoviscosity, 36, 60, 79, 84, 86, 88–90, 92, 94, 128, 142, 164, 197, 208
piston, 1–4, 6, 8, 9, 10, 12, 14, 20, 21, 23, 24, 26, 31, 35, 40, 61, 82, 84, 92, 123–125, 127, 128, 133–136, 139–158, 180, 205
  bearing, 133, 135, 136, 141, 142, 144–149, 154–157, 158
power
  law, 35, 128, 142
  loss, 85, 93–96, 107, 110, 111, 113, 201, 211, 212, 215, 220
pressure field, 8, 22, 66, 72–75, 78, 86, 91, 92, 95, 97, 133, 136, 137, 149, 150, 155–157, 165, 172, 174, 182, 186, 191, 192

R, S, T

radial
  clearance, 35, 60, 76, 77, 84, 102, 106, 128, 142, 144, 164, 175, 181, 201, 209, 210, 211
  shape, 209

Reynolds number, 108
screening, 200, 201
supply groove, 165
TEHD problem, 42, 44, 45
temperature field, 67, 85, 103, 104, 106, 114
thermal
  boundary layer, 104
  conductivity, 102, 106
  deformation, 102, 103, 171, 175
  model, 105, 110, 116, 118, 175, 205, 216
transfer coefficient, 101, 105–115

U, V, W

unilateral contact, 39–42, 46
V engine, 72, 76, 77, 78
wear field, 97, 98
weighting function, 138